Spatial Politics

RGS-IBG Book Series

Published

Forthcoming

Spatial Politics

Essays for Doreen Massey

Edited by

David Featherstone
and Joe Painter

WILEY-BLACKWELL

A John Wiley & Sons, Ltd., Publication

This edition first published 2013
© 2013 John Wiley & Sons, Ltd

Wiley-Blackwell is an imprint of John Wiley & Sons, formed by the merger of Wiley's global Scientific, Technical and Medical business with Blackwell Publishing.

Registered Office
John Wiley & Sons, Ltd, The Atrium, Southern Gate, Chichester, West Sussex, PO19 8SQ, UK

Editorial Offices
350 Main Street, Malden, MA 02148-5020, USA
9600 Garsington Road, Oxford, OX4 2DQ, UK
The Atrium, Southern Gate, Chichester, West Sussex, PO19 8SQ, UK

For details of our global editorial offices, for customer services, and for information about how to apply for permission to reuse the copyright material in this book please see our website at www.wiley.com/wiley-blackwell.

The right of David Featherstone and Joe Painter to be identified as the authors of the editorial material in this work has been asserted in accordance with the UK Copyright, Designs and Patents Act 1988.

Library of Congress Cataloging-in-Publication Data

Spatial Politics: Essays for Doreen Massey / edited by David Featherstone and Joe Painter.
    p. cm.
  Includes bibliographical references and index.
  ISBN 978-1-4443-3831-7 (cloth) – ISBN 978-1-4443-3830-0 (pbk.)   1. Political geography.   2. Massey, Doreen B.   I. Massey, Doreen B.   II. Featherstone, David, 1974–
III. Painter, Joe, 1965–
  JC319.S616 2013
  320.12–dc23

                                                    2012025404
A catalogue record for this book is available from the British Library.

Cover image: © Ingrid Pollard
Cover design by Workhaus

Set in 10/12pt Plantin by SPi Publisher Services, Pondicherry, India

1   2013

Contents

List of Figures

Cover: Quandan is one of a group of photographs which Ingrid Pollard exchanged with Doreen Massey after hearing her give a lecture at the Royal Institute of British Architects. Through the photographs, Ingrid and Doreen began an ongoing conversation about everything from space to geology. The images relish their duplicity in the developments of the rules of aesthetics, of astronomy, surveying and mapping, and geometry. Together they combine to produce a sense of wonder.

Ingrid Pollard

Doreen Massey and the editors of the collection offer many thanks to Ingrid for her permission to use this image for the cover.

Notes on Contributors

Ash Amin is the 1931 Chair in Geography at the University of Cambridge. Recent publications include *Land of Strangers* (Polity Press, 2012) and, with Nigel Thrift, *The Politics of World-Making* (Duke University Press, 2013).

Sophie Bond is currently a lecturer in Environmental Studies and Geography at Victoria University, Wellington in Aotearoa New Zealand. She is interested in theories of radical democracy and how these might be mobilised to create spaces of contestation and opportunities for local mobilisation that can create alternative futures to contest the dominance of the apolitical neoliberal present.

Allan Cochrane is Professor of Urban Studies at the Open University. His research interests lie at the junction of geography and social policy, and he has researched and published on a wide range of topics relating to urban and regional policy and politics. He was co-author (with John Allen and Doreen Massey) of *Re-Thinking the Region* (Routledge, 1998). His book *Understanding Urban Policy: A Critical Approach* was published by Blackwell in 2007.

Andrew Cumbers is Professor in Geographical Political Economy at the University of Glasgow. Recent publications include *Reclaiming Public Ownership: Making Space for Economic Democracy* (Zed Books, 2012).

Elena dell'Agnese is Chair of the Commission of Political Geography of the IGU, and teaches Political Geography at the University of Milano-Bicocca. Her main research interests are Political Geography, Gender and Media Studies. She published *Geografia Politica Critica* in 2005, and *Paesaggi ed eroi. Cinema, nazione, geopolitica* in 2009.

Olafur Eliasson, born in 1967, represented Denmark at the 50th Venice Biennale in 2003 and later that year installed *The weather project* at Tate Modern. *Take your time: Olafur Eliasson*, a survey exhibition organised by

SFMOMA in 2007, travelled until 2010, to the Museum of Modern Art, among other locations. *Seu corpo da obra* (Your Body of Work), which opened in September 2011, occupies three different venues in São Paulo and extends into the city itself. *Your rainbow panorama*, a 150-metre circular, coloured-glass walkway on top of the museum ARoS in Aarhus, Denmark, opened in May 2011. The facade for Harpa Reykjavik Concert Hall and Conference Centre, inaugurated in August 2011, was created by Eliasson in collaboration with Henning Larsen Architects. Established in 1995, his Berlin studio today numbers about 45 craftsmen, architects, geometers and art historians. In April 2009, as a professor at the Berlin University of the Arts, Olafur Eliasson founded the Institut für Raumexperimente (Institute for Spatial Experiments).

Arturo Escobar is Distinguished Kenan Professor in Anthropology at the University of North Carolina, Chapel Hill. His most recent book is *Territories of Difference: Place, Movements, Life, Redes* (Duke University Press, 2008). He co-edits with Dianne Rocheleau the book series 'New Ecologies for the Twenty-First Century' for Duke University Press.

David Featherstone is Senior Lecturer in Human Geography at the University of Glasgow. He has key research interests in space, politics and resistance in both the past and present. He is the author of *Resistance, Space and Political Identities: The Making of Counter-Global Networks* (Wiley-Blackwell, 2008) and *Solidarity: Hidden Histories and Geographies of Internationalism* (Zed Books, 2012).

Lawrence Grossberg is Morris Davis Distinguished Professor of Communication Studies and Cultural Studies at the University of North Carolina. His recent publications include *Caught in the Crossfire: Kids, Politics and America's Future* (Paradigm, 2005) and *Cultural Studies in the Future Tense* (Duke University Press, 2011).

Rogério Haesbaert is Professor in Geography in the Instituto de Geociências, Universidade Federal Fluminense, Brazil. He has published extensively on political geography and multi-territoriality, including a book translated from Portuguese to Spanish, *El mito de la Desterritorialización* (Siglo Veintiuno Editores, 2011).

Wendy Harcourt is senior lecturer in social policy at the International Institute of Social Studies of the Erasmus University, The Hague, and editor of the quarterly journal *Development*. She received the 2010 Feminist and Women's Studies Association's Prize for her book *Body Politics in Development: Critical Debates in Gender and Development* (2009). Her fifth edited collection, *Women Reclaiming Sustainable Livelihoods: Spaces Lost, Spaces Gained* (Palgrave) was published in April 2012, and she is completing a monograph titled *Gender and Environment: An Introduction* for Zed Books. She is editor of two book series, 'Gender and Environment' (Zed Books) and 'Gender,

Development and Social Change' (Palgrave), and co-editor of the *International Handbook on Transnational Feminism* for OUPA.

Steve Hinchliffe is Professor in Human Geography at Exeter University in the UK. Before this he was reader in Environmental Geography at the Open University, enjoying the constant intellectual challenge of working with Doreen Massey. He has written and edited a number of books and journal special editions, including *Geographies of Nature* (Sage, 2007). He is currently working on an ESRC-funded project entitled 'Biosecurity Borderlands', and on a European Union-funded 'Science in Society' project which develops research collaborations with civil society organisations on environmental problems in Europe.

Sara Kindon lectures in Human Geography and Development Studies in the School of Geography, Environment and Earth Sciences, Victoria University of Wellington, Aotearoa New Zealand. She has published on participatory action research and the use of participatory video in geography. She is particularly interested in furthering the development of decolonising research practices as part of her ongoing work with Maori and refugee-background research partners.

Ken Livingstone was leader of the Greater London Council from 1981 until its abolition in 1986, and Mayor of London between 2000 and 2008. He was MP for Brent East from 1987 to 2000.

Ricardo Menéndez is the Minister of Science, Technology and Intermediate Industry in the Venezuelan government. Before this he was Professor of Geography at the Instituto de Urbanismo, Universidad Central de Venezuela in Caracas.

Chantal Mouffe is Professor of Political Theory at the University of Westminster. She is co-author with Ernesto Laclau of *Hegemony and Socialist Strategy: Toward a Radical Democracy* (Verso, 1985.) Her most recent book is *On The Political* (Routledge, 2005).

Joe Painter is Professor of Geography at Durham University. He mainly works in the fields of political and urban geography. His current research interests include the geographies of the state and governance, questions of citizenship and democracy, geographies of work and play, and theories of region and territory. He has published widely on these and other topics and is the co-author of *Political Geography* (Sage, 2009) and of *Practising Human Geography* (Sage, 2004).

Jamie Peck is Canada Research Chair in Urban & Regional Political Economy and Professor of Geography at the University of British Colombia. A recipient of Guggenheim and Harkness fellowships, Peck has previously taught at the University of Manchester and the University of Wisconsin-Madison. He has research interests in the historical geographies of neoliberalism, labour geography and urban political economy. Recent publications include *Constructions of Neoliberal Reason* (Oxford University Press, 2010) and *The Wiley-Blackwell Companion to Economic Geography* (Wiley-Blackwell, 2012, co-edited with Trevor Barnes and Eric Sheppard).

Dianne Rocheleau is Professor of Geography at the Graduate School of Geography, Clark University. Her interests include environment and development, political ecology, forestry, agriculture and landscape change, with an emphasis on the role of gender, class and 'popular' versus 'formal' science in resource allocation and land use. She is currently completing a book manuscript, *The Invisible Ecologies of Machakos: Landscape, Livelihoods, and Life Stories 1890–1990*. She co-edits with Arturo Escobar the book series 'New Ecologies for the Twenty-First Century'.

Paul Routledge is a Reader in Human Geography at the University of Glasgow. Recent publications include, with Andrew Cumbers, *Global Justice Networks: Geographies of Transnational Solidarity* (Manchester University Press, 2009).

Michael Rustin is Professor of Sociology at the University of East London and a Visiting Professor at the Tavistock Clinic. He was a co-founder of *Soundings* with Stuart Hall and Doreen Massey. His recent publications include (with Margaret Rustin) *Mirror to Nature: Drama, Psychoanalysis and Society* (Karnac Books, 2002).

Arun Saldanha is an Associate Professor at the University of Minnesota. His recent publications include *Psychedelic White: Goa Trance and the Viscosity of Race* (University of Minnesota Press, 2007).

David Slater is Emeritus Professor of Political Geography at Loughborough University, and Associate Fellow of the Institute for the Study of the Americas, University of London. He is the author of *Geopolitics and the Post-Colonial*, and of a number of recent articles on imperial power, including 'Rethinking the imperial difference' in *Third World Quarterly* (2010) and 'Latin America and the challenge to imperial reason' in *Cultural Studies* (2011).

Nigel Thrift is Vice Chancellor of the University of Warwick. He is the author of numerous influential books and papers. His most recent publications include (with P. Glennie) *Shaping the Day* (Oxford University Press, 2009) and *Non-Representational Theory: Space, Politics, Affect* (Routledge, 2007).

Hilary Wainwright is Research Director of the New Politics Programme at the Transnational Institute and the editor of *Red Pepper*. Her recent publications include *Reclaim the State: Adventures in Popular Democracy* (Verso, 2003).

Jane Wills is Professor of Human Geography at Queen Mary, University of London. Her recent publications include (with Kavita Datta, Yara Evans, Joanna Herbert, Jon May and Cathy McIlwaine) *Global Cities at Work: New Migrant Divisions of Labour* (Pluto, 2009); (with Angela Hale) *Threads of Labour: Garment Industry Supply Chains From the Workers' Perspective* (Blackwell, 2005); (with Peter Waterman) *Place, Space and the New Labour Internationalisms* (Blackwell, 2001).

Alice Brooke Wilson is an anthropology graduate student at the University of North Carolina, Chapel Hill, studying food politics and social movements in Mexico and the United States. She is also co-founder of Maverick Farms, an organic farm dedicated to food justice and education in the Appalachian Mountains (www.maverickfarms.com).

Foreword

Ken Livingstone

I am delighted to be writing the foreword to this collection recognising what Doreen Massey has done, especially given the importance to me of the connections we've had since we first met in 1976. I can remember the first time I saw Doreen. She was standing, on a cold November evening, outside the Tenant's Hut in Kilburn in London where we were due to have our Labour Party ward meeting, as I had just been selected as the Labour candidate. It was an absolutely miserable night; there were about three of us there. I think she was the first professional geographer I'd ever met. And I'd never had any desire to meet any, because when I did O level geography (one of the few O levels I got before I dropped out of school), it mainly consisted of drawing maps, and remembering heights and mountains and rivers. As far as it ever got political was when our teacher told us the role that climate and coastline had played in holding back indigenous cultures from reaching the excellence of the British Empire. In fact, as I recall we only focused on those countries that had been part of the British Empire, though by the time I was at school most of them had escaped from it.

In the mid-1970s the debate on the left in Britain was between Stuart Holland's Alternative Economic Strategy (*The Socialist Challenge*, 1975) and the wholesale nationalisation proposed by the far left. Against the background of North Sea oil to fund infrastructure and the modernisation of industry, everything seemed possible. Contrary to those who rewrote history to depict the 1970s as an ungovernable decade, this was the high point of the post-war social democratic settlement. The top rate of tax was down from 98 per cent after the war to just 80 per cent, but with death duties we had lived through 30 years of redistribution of wealth with the top 10 per cent earning just four times the bottom 10 per cent.

The strength of the trade unions was a restraint on excessive corporate pay and bonuses; the working day was cut by 40 minutes during that decade; holiday entitlement doubled; and women's pay dramatically closed the gap on men's. I believed we were on an irreversible journey to a socialist society, viewing Margaret Thatcher and her Friedmanite beliefs as a throwback to pre-Keynesian times. The idea that Thatcher, Reagan and their heirs would over 30 years wind the world back to levels of inequality not seen since the First World War was inconceivable. Even if Labour's faltering leadership opened the way to a Thatcher government, I had no doubt the left would capture the party and return to power after one Thatcher term.

When I became leader of the Greater London Council (GLC) in 1981, Doreen was one of the first people I turned to for some input into our industry and employment policy. She was one of my appointees to the Greater London Enterprise Board which we set up to try to really analyse and then correct all the things that Doreen still thinks are wrong with the London economy. And they weren't quite as bad then as they are now. We didn't make a lot of progress on turning that round, but it wasn't our fault because an evil tyranny abolished us.[1]

It was at about that time that Doreen got a job at the Open University, and that was very interesting. After she went for the interview, months had passed and they never announced who had been offered the job. There was discussion about why so much time had passed; people talked about whether she would have sufficient 'gravitas' for the post. Given that she had spiky multi-coloured hair at the time, she didn't look like the typical professor they were used to. Then, shortly after it had been confirmed that she'd got the job, Margaret Thatcher's Education Secretary, Keith Joseph, was wandering round another university and when he was introduced to the geography department there he said he was very worried about the coming politicisation of geography. That made us did think there might have been a political undercurrent to the delay.

After the abolition of the GLC in 1986 we decided it was worth carrying on the debates and the political project it had been part of. We formed the Ariel Road group – Ariel Road is where Doreen lives – and had very intense debates over many years about what was happening in this post-Fordist world and what we should do politically. Then, when I became Mayor of London in 2000, my relationship with Doreen became a bit strained. She was pounding on with her examination about what was rotten at the heart of the London economy and what it does to the rest of the world – and I was Mayor of the city. On one occasion she turned up to interview me and I felt she was about to leap over and bang my head on the table to make her point that 'you've got to do more to change it'.

Such are the dynamics and contradictions of politics. I always used to say, when we were discussing the London economy, that this is not the world I would have created, it's the world we're stuck with. But not any longer, perhaps; and this is where our opportunities come today, because the scale of what has happened in the recent financial crisis dwarfs anything since the Great Depression.

So this book appears at a perfect point to reassess the past, with the post-war period dividing neatly into three social democratic decades and three neoliberal ones. Contrary to the right's complaints that the public sector crowded out private investment and their promise that increased inequality of wealth would have a trickle-down effect, in Britain average incomes grew by 2.4 per cent per year in the 1960s and 1970s but dropped to 1.7 per cent per year in the last 20 years. In the world at large, growth during the social democratic era averaged 3 per cent a year, but was reduced to half that rate in the last 30 years.

Crushing the trade unions did not lead to a revival of our economy, but allowed a shift to short termism as manufacturing was wiped out and replaced by the growth of finance, which no longer provided funds for investment. Instead, almost every aspect of the economy was turned into an opportunity for speculation – or to be more honest, gambling. A Britain that once led the world in exporting manufactured goods now had the dubious honour of being the world's hedge fund capital.

As the public utilities were privatised the public faced a huge increase in prices. Hundreds of thousands of skilled working-class jobs were eliminated whilst the utility bosses paid themselves vast increases in salary. Nowhere are the consequences of privatisation clearer than in the building societies. Not a single building society that demutualised remains as an independent institution. The New Labour government's privatisations have been equally unsuccessful, with the catastrophic waste of billions of pounds on Tube privatisation and with the NHS now crippled by the obscene costs of funding the private finance initiative (PFI) scams.

Any objective person looking at these two periods cannot fail to recognise that the balance between the public and private sectors and the redistribution of wealth in the immediate post-war period produced not just a fairer society but a more economically successful one. Yet as the Thatcher/Reagan era imploded throughout 2008 and governments of all colours rushed to prevent the collapse of the banking system, creating a depression as bad as that of the 1930s, the right in both Britain and the USA seized the opportunity to make the case for a yet smaller state.

With breathtaking dishonesty (unchallenged by the bulk of the media), Prime Minister David Cameron and Chancellor of the Exchequer George Osborne justify their cuts by claiming that Britain now has the largest government deficit in British history, even though in real terms it is just a third of the size of that faced by the 1945 Labour government and, unlike the structure of Greek government debt, has an average repayment date 14 years in the future. Ireland is further down the road now taken by Cameron and Osborne. Ireland's savage public sector cuts have actually seen the bond markets *increase* the cost of loans whilst the deficit has increased because workers thrown on the scrap heap are neither paying taxes nor generating GDP. The danger of our government's cuts is that they will either push Britain back into a recession or leave us limping along for a decade or more with Japanese levels of low growth.

The success of the Cameron government is to have won the 'spin' battle over the deficit. Instead of massive public anger that bankers are once again lining their pockets whilst the majority of society bears the pain of their folly, the bulk of the public has been persuaded that we are in this mess because of high levels of public spending and 'gold-plated' public sector pensions. Those of us who believe in a different strategy are presently undermined by the complicity of some senior Labour politicians in supporting the Tory claim that deficit reduction must take priority over investment. The most significant speech by any politician in the months after the general election in May 2010 was Ed Balls's address to journalists at Bloomberg on 27 August 2010. Drawing on the lessons of history, Balls systematically demolished the Tory case and laid out the alternative strategy of investment-generated growth. That speech provided the arguments for everyone seeking to challenge the government strategy and lays the foundation for the next Labour government to avoid repeating the mistakes of the Blair/ Brown years.

In a way, the most surprising thing is that such a strategy did not underpin the last Labour government – the link between investment and economic success is clear throughout the entire period of modern capitalism. Britain achieved domination of the nineteenth century because it became the first nation in history to invest 7 per cent of its GDP. Following the American Civil War, an investment rate in the high teens guaranteed that the twentieth century would be American. After the Second World War, West Germany led the rest of Western Europe in a growth spurt which in the early 1970s saw the Germans investing 25 per cent of their GDP (Britain always lagged about 5% behind). Japan leapt from being a post-war bombsite to being the world's second largest economy by 1968 by driving investment up to 38 per cent of GDP in its peak year.

The rest of Asia watched and learned. In India investment limped from 10 per cent of GDP at Independence to just 20 per cent in 2004 when the Singh government was elected. He oversaw a surge to 35 per cent and was overwhelmingly re-elected five years later. But the great investment success story is, of course, China. Thirty years after Deng Xiao Ping initiated the new economic policy, China has overtaken Japan as the second largest economy on Earth. Unlike the West, where the response to the banking crisis was a dramatic fall in investment, in China the rate was increased from 43 to 46 per cent. At this rate, ignoring the distortions of the exchange rate, in real terms China will become the world's largest economy during the present decade.

As we cast about to find a way out of our economic problems, any strategy that ignores the lessons of investment is doomed to fail. Investment in infrastructure, plant and the education of the workforce is the key – and never more so. For 500 years the European empires and then America used their military power to rig the world economy in their interests, allowing us to live a lifestyle based on the exploitation of the majority of humanity. That period is closing. In devising a new strategy for our economy it has to be placed in the context of the new world that is being born in Asia, Africa and Latin America.

Given such geographical upheavals, there is a huge desire to rethink where we are, examine the problems we face and to come up with solutions. Doreen's political and theoretical work over the past 40 years has been driven by attempts to forge such alternatives. As we analyse the forces coming into play, the work of Doreen Massey will remain indispensable.

Note

1   This is a reference to the abolition of the Greater London Council by Margaret Thatcher's government in 1986. As Hilary Wainwright argues, in her chapter in this collection, 'in most countries, the destruction of a level of government, against the will of the majority of citizens, is associated with authoritarianism verging on dictatorship'.

Reference

Holland, S. (1975) *The Socialist Challenge*. London: Quartet Books.

Series Editors' Preface

The RGS-IBG Book Series only publishes work of the highest international standing. Its emphasis is on distinctive new developments in human and physical geography, although it is also open to contributions from cognate disciplines whose interests overlap with those of geographers. The series places strong emphasis on theoretically informed and empirically strong texts. Reflecting the vibrant and diverse theoretical and empirical agendas that characterise the contemporary discipline, contributions are expected to inform, challenge and stimulate the reader. Overall, the RGS-IBG Book Series seeks to promote scholarly publications that leave an intellectual mark and change the way readers think about particular issues, methods or theories.

For details on how to submit a proposal please visit:
www.rgsbookseries.com

Neil Coe
National University of Singapore

Joanna Bullard
Loughborough University, UK

RGS-IBG Book Series Editors

Acknowledgements

This book was instigated by John Allen, whose sage advice and gentle encouragement throughout the project have been indispensable. Steve Pile has offered very useful guidance, particularly in the final stages of the book. The Open University geography department in general has offered much help throughout the process and we would particularly like to thank Allan Cochrane, Gillian Rose and Jenny Robinson (now at UCL).

The Series Editors for the RGS-IBG, Kevin Ward and then Neil Coe, have been enthusiastic and have provided generous editorial engagement. Jacqueline Scott at Wiley-Blackwell has been a very supportive (and patient!) editor. We would like to thank an anonymous reader for their thoughtful and constructive comments on an earlier version of the manuscript. Robin Jamieson and Cheryl McGeachan both provided invaluable help with formatting and bibliographic work.

Dave Featherstone would like to thank Neil Gray, Hayden Lorimer, Danny MacKinnon, Richard Phillips and Chris Philo, who all made characteristically wise suggestions at crucial junctures of the project. Mo Hume's help and support has been invaluable as ever, particularly in liaising with Ricardo Menéndez. Aoibhe and Marni have made the whole process much more enjoyable.

Joe Painter would mainly like to thank Dave Featherstone for his patience, as well as Rachel Woodward, Ruth and Patrick for their love and support.

We would both like to thank all the contributors for their enthusiasm, patience and engagement.

Most of all, we would like to thank Doreen herself for her enthusiasm and passion for the project. She has been generously engaged with what at times must have seemed a mysterious book and has answered many a query as the project has progressed. We hope she likes the end result and that there are not too many stray apostrophes.

David Featherstone and Joe Painter
Glasgow and Durham, May 2012

The Lipman–Miliband Trust

Royalties from this book will be donated to the Lipman–Miliband Trust. This Trust, on which Doreen has served as a trustee for many years, exists to support socialist education and research. For more details see: www.lipman-miliband.org.uk/

Introduction
'There is no point of departure': The Many Trajectories of Doreen Massey

David Featherstone and Joe Painter

In late October of 2011 Doreen Massey addressed the Occupy London encampment outside St Paul's Cathedral in the heart of London's financial district. Part of the transnational movement seeking to shape a future beyond aggressive financial capitalism, Occupy London had set up camp in the City on 15 October. Massey attacked the 'invasion of the imagination' that has defined political reaction to the financial crisis.[1] Contending that politicians are 'scared' of the financial centre and its power, she asserted the importance of the camp in challenging the social relations of the City. For Massey, such challenges to the 'current construction and role of a place' are integral to forging alternative political futures (Massey, 2004: 17).

Doreen Massey's compelling contributions to geographical theorising and political debate have been animated by such insights and political commitments. Her official retirement from the Open University in September 2009 has in no sense slowed the dynamism of her political and theoretical work. It provided, however, an opportunity to reflect and take stock of her diverse contributions. This collection brings together former graduate students, colleagues, geographers and other social scientists with artists, political figures and activists. It seeks to honour, engage with and take forward Doreen Massey's vital geographical and political contributions.

Introducing the many trajectories and engagements of Doreen Massey is no mean challenge. Her whole theoretical approach is antithetical to the ways such accounts are often structured around paradigm shifts which position disciplines in relation to sequential narratives or clashes between key disciplinary figures. Any attempt to fit her work into a neatly sequential account of geography's recent past, with perhaps a tidy temporalising of difference between Marxism, feminism and post-structuralism, would be doomed to fail. Doreen Massey's own ethos is contrary to these ways of thinking about politics and intellectual work. Her liveliness and openness to the world exceeds such boundaries and confining categories. In what may have been a precocious commitment to spatial particularity, she is one of a very few Mancunians who have been lifelong supporters of Liverpool Football Club. The intense concentration of a supervision meeting with Doreen might be broken as she paused to wonder whether it was a white-throat that had just flown past the window.

This introduction engages with the multiple trajectories at work in Massey's writing and in her political interventions across different times and spaces. We explore six coeval influences which have shaped her work and which she has in turn shaped; Althusserian Marxism, feminism/ difference, internationalism, class and inequality, materialities and thinking spatially. These influences cross-cut the four key sections of the book: 'Space, Politics and Radical Democracy', 'Regions, Labour and Uneven Development', 'Reconceptualising Place' and 'Political Trajectories'. In the spirit of Massey's work the collection seeks to be alive to the unexpected possibilities shaped through bringing together trajectories, perspectives and experiences from diverse contexts and positions.

Space, Politics and Radical Democracy

Doreen Massey begins her book *For Space* by remarking that she has been 'thinking about "space" for a long time'. She continues that 'usually I've come at it indirectly, through some other kind of engagement. The battles over globalisation, the politics of place, the question of regional inequality, the engagements with "nature" as I walk the hills, the complexities of cities' (Massey, 2005: 1). Massey's engagements with space have never been conceived only as an intellectual end, however passionately committed she is to a geographical approach to the world. Rather, she mobilises a spatial perspective as a set of intellectual and political tools to bring analytical clarity and purchase on diverse situations. It is because of the traction that such tools give to political questions that they have been so central to her project.

It is in the process of wrestling over such political questions as regional inequality and globalisation that Massey has thoroughly reworked many key assumptions about space and place. In doing so she has profoundly reshaped common-sense ways of thinking about space and place both in the discipline of geography and across the broader social sciences. Her work has provoked a significant rethinking of space as the product of relations rather than as a fixed surface or container. She has stressed that space is the dimension of multiplicity, is the product of relations and is always unfinished and under construction. In opposition to long-standing assumptions that time is what gives politics life she has insisted that taking spatiality seriously has important implications and it cannot be rendered as secondary to temporality. To do so, she has consistently argued, is to greatly impoverish our political imaginations.

Many of Massey's insights are now a familiar part of the terrain of geographical thinking. In this regard it is sometimes easy to forget the intellectual labour and struggle it took to get these insights recognised as central and significant. The idea that space and politics are co-constitutive, that they are built together as the outcomes of different ongoing processes, for example, is now commonplace in accounts of the relations between geography and the political. This position was, however, articulated over a long duration, through many different iterations, and in the face of serious opposition to the idea that space should be accorded political status. Her position was forged through argument with influential political theorists including Ernesto Laclau.

Massey's challenge to Laclau's contention in *New Reflections on the Revolutions of Our Time* that politics and space are 'antinomic terms' was a particularly significant intervention. Laclau contended that 'Politics only exist insofar as the spatial eludes us' (Laclau, 1990: 68). Massey used a critique of Laclau's position as the starting point of her essay 'Politics and Space/Time', which insists on the co-production of space, time and politics. As she argues, 'the spatial is integral to the production of history, and thus to the possibility of politics, just as the temporal is to geography' (Massey, 1992: 84). This assertion of the spatio-temporal construction of politics opens up a focus on the diverse, multiple and contested processes of the political. The essay notes significant parallels between her position and developments in Einsteinian and post-Einsteinian theories of time–space relativity, noting that in 'modern physics' the identity of things 'is *constituted through* interactions' (Massey, 1992: 76, emphasis in original).

Massey has consistently stressed that different ways of thinking space have consequences and effects. For Massey, to think spatially is never an innocent, politically neutral activity. Rather, ways of conceptualising space have important effects and consequences. Whether it is a Conservative politician arguing that the Taliban are medieval rather than contemporaneous, or the stage-ist conceptions of development that position some places as 'ahead' of others, Massey has insisted that spatial imaginations have political consequences and effects. Similarly, she has consistently argued that the way in which some regions are positioned unequally needs to be seen not as natural or inevitable, but part of relations of power (Massey, 1978). This has been part of a broader reinvigoration of ways of thinking space, particularly associated with critical geography. In these debates Massey's voice has persistently been distinguished by its clarity and originality.[2]

The contributions of Massey to thinking spatially present important challenges to dominant ways of thinking about politics and the political. As Chantal Mouffe argues in her contribution here, Massey's assertion that space is the dimension of multiplicity is fundamental to the question of how we are to live together. It is central to understanding the ways in which different social relations can be envisioned and different forms of political agency constructed. Massey insists that as space is always under construction and never finished, this means that it is always possible that spatial relations can be articulated and generated in different, potentially antagonistic, ways.

Massey's commitment to this reworking of space has a strong Althusserian influence and lineage. That she came to Marxism via Althusser makes her trajectory distinctive. The generation of radical geographers who emerged in the 1970s were largely hostile to (or didn't read) Althusser (Philo, 2008: xxxvii–xxxviii; see also Duncan and Ley, 1982). Indeed it is remarkable, certainly by comparison with cognate social science disciplines, how strongly Marxist geography has been constructed largely from classical Marxist sources. David Harvey's treatise(s) on capital are clearly the most significant in this regard (see Harvey, [1982], 2007). Henri Lefebvre, whose work was explicitly positioned in opposition to structuralism, was also a significant influence on Harvey and Neil Smith, though he was also to influence Doreen's work (see Lefebvre, 2006).

Reading Althusser, in the rather unlikely place of Pennsylvania, transformed Massey's attitude to Marxism. What so excited her about Althusser was the way that his thinking transcended the humanism of the early Marx. As a feminist she was deeply suspicious of the gendered assumptions of such humanism. Thus she argues that:

> Far from liberation, it seemed a trap. It was Althusser, with his ideas of structured causality (most especially 'Contradiction and Overdetermination' and his famous dictum that 'there is no point of departure', in other words on my reading there is nothing which you have to accept as eternally pre-given, which is not in itself a product of previous causal structures) who provided a welcome way out. (Massey, 1995b: 351; see also the interview below).

Massey's account here challenges ways in which accounts of the emergence of radical geography have frequently positioned Marxism and feminism as oppositional or sequential, rather than coexisting influences (Rose, 1993: 113). Feminism and Marxism here emerge as coeval trajectories in Massey's work. She emphasises the liberating importance of Althusser's argument that 'there is no point of departure' with its implications that 'nothing is given' and therefore everything can be challenged and reworked (see Althusser, 1971: 85). She has argued that 'as a young woman who was trying to escape the norms, who didn't conform to any of the given descriptions of "woman", and who wanted a way of challenging them, that first entry into anti-essentialism, although I didn't know that term, none of us knew that term at that point, was utterly important' (Massey *et al.*, 2009: 404).

Stuart Hall, Massey's colleague at the Open University and co-founder of the journal *Soundings*, suggests that Althusser's 'non-teleological reading of Marx' led to a concern with how different structures were articulated together, rather than seen as temporally discrete stages 'with a necessary progression built into them' (Hall, 1980: 326; see also Althusser, 2005: 89–128). As Arun Saldanha argues here, Massey extracted from Althusserian Marxism 'an appreciation of the open temporality of social formations and the multiple relationships intersecting in those formations'. This allows a focus on the 'enablement of new relations', such new relations being generated through the coming together of previously separate or disparate trajectories (see Massey, 2005: 40).

During the 1980s the terrain of Althusserian Marxism was reworked and rearticulated through Ernesto Laclau and Chantal Mouffe's contentious and influential work on radical democracy. Their book *Hegemony and Socialist Strategy* mobilised Althusserian concepts like articulation as part of a plural reimagining of left political strategies. Many left figures and theorists directly attacked Laclau and Mouffe's position for disrupting left certainties and condemned it as a retreat from class politics (see for example Geras, 1987). Massey, however, embraced the possibilities opened up by thinking about the plural construction of left politics. This certainly was not out of anything like a disavowal of left commitments. As Michael Rustin argues here, while 'there is a "post-modern" cast to Massey's ontological and epistemological positions (the ontology is somewhat constructivist, the epistemology somewhat pragmatist) her substantial political and ethical commitments have remained unyielding'. She also has argued for the importance of 'thinking radical democracy spatially' in ways which usefully ground the rather abstract ways of thinking the political that structure Laclau and Mouffe's work (Massey, 1995a). The conversation between her work and particularly the work of Chantal Mouffe has been mutually productive.

Mouffe's chapter here mobilises Massey's account of the 'power-geometries of globalisation' to critique Hardt and Negri's use of the concept of 'smooth space' to analyse the forms of global political activity that they term 'the multitude' (see Hardt and Negri, 2001). Massey's critique of the political framings of globalisation is explored in the chapters by Lawrence Grossberg and David Slater. Massey's work from the early 1990s has been increasingly concerned with interrogating the 'power-geometries of globalisation' (Massey, 1993a). In the face of dominant political and theoretical discourses which have constructed globalisation as an inevitable force, Massey's analysis has insisted on the need to think about globalisation as an uneven, contested and riven process (see especially 1999a, 2005). She has powerfully argued that the spatial imaginaries at work in globalisation discourses, whether they be the assumption that globalisation is Westernisation, the sense that globalisation results in the 'annihilation of space by time', or homogeneity of place are not innocent. Rather, Massey contends that they do significant work in furthering the political projects of (neoliberal) globalisation. As Grossberg notes here, 'If the past, the other, was never as simple, or homogeneous or local or unified as we imagine, the present is probably not as fractured or heterogeneous or global as we assume.'

Massey has been insistent on the importance of understanding globalisation as a neoliberal project. This permits a focus on the analysis of the mechanisms and practices that globalisation has been formatted and constituted through neoliberal practices and conventions. The political resonances and potential of this position are also important. Some opposition to globalisation has fallen into the political trap of counter-posing the local and global in problematic ways. Massey's insistence that current globalisation is part of a neoliberal project invites the possibility of imagining alternative ways of generating globalisation (see also Featherstone, 2008). Her work, consistent with her internationalist political convictions, has insisted on the importance of and possibilities for reworking and reimagining practices of internationalism in 'global times'. David Slater uses this approach as a starting point for a post-colonial account of the relations between 'space, democracy and difference'. He critically reflects on the relations between radical democracy, spatiality and politics and the importance of thinking about the generative character of West/non-West interactions. This commitment to a relational construction of space and politics has also been forged through an important challenge to dominant ways of thinking about regional politics and geographies.

Regions, Labour and Uneven Development

Much of Doreen Massey's early work was concerned with regions, labour and uneven development. During the 1970s she worked at the Centre for Environmental Studies (CES), an independent research centre funded by the UK government and the Ford Foundation. It acquired a reputation for radicalism and, following Margaret Thatcher's first election victory, it was closed as part of government spending cuts. During her time at CES (which included a period of graduate study at the University of Pennsylvania in the United States) Massey developed a critique of the then dominant form of industrial location theory and of its grounding in assumptions drawn from neoclassical economics. This critique led to her elaboration of the alternative approach for which she has become particularly well known and which was set out in her path-breaking book, *Spatial Divisions of Labour* (Massey, 1984).

In the early 1970s, Massey went the United States to study mathematical economics and regional science – subjects that epitomised the then orthodox neoclassical approaches to modelling the geography of economic activity (or economic 'behaviour', as the terminology of the time might have put it). As noted earlier, it was during her stay in Pennsylvania that she encountered the ideas of Louis Althusser. In 1973 Massey published a paper in *Antipode*, with the title 'Towards a critique of industrial location theory', in which she took classical industrial location theory to task for its ideological character, its failure to situate its object of study in historical context and its lack of attention to the specifically capitalist character of the economic system (Massey, 1973). Neither Althusser nor Marx is mentioned, but the Althusserian influence is apparent in both the substance and the mode of its arguments (Swyngedouw, 2000). This intervention was one of a number made by geographers (often in the pages of *Antipode*) who were influenced by Marx and other critical thinkers and who questioned the asocial and ahistorical approach to geography known as spatial science.

In Massey's case, critique was soon followed by the development of alternative approaches, including research on capitalist land ownership (Massey and Catalano, 1978) and the introduction of the idea of 'spatial divisions of labour' (Massey, 1978, 1979, 1984). The affinities between the approach taken in *Spatial Divisions of Labour* and critical realism have been widely noted. The book's account of the complexity of regions also echoes Althusser's concepts of overdetermination and articulation (Althusser, 2005). Althusser argued that the contradiction between capital and labour never appears in pure form, but is always 'overdetermined' by the specific characteristics of the situation. Regions can also be seen as the products of overdetermination: successive 'rounds of investment' are combined and interact with the distinctive landscapes of class and political relations that have arisen from preceding rounds. The results are regional geographies that are unique and cannot be explained in terms of the working out of a unitary logic, but which can nonetheless be understood as the products of multiple determinations. As Massey makes clear, this is quite different from determinism: 'the aim is conceptualisation and understanding, not "prediction" of pre-determined outcomes' (Massey, 1995b: 305).

The concept of spatial divisions of labour enabled Massey to develop a theoretically rigorous understanding of regional uneven development that refused to take regions as pre-given entities: 'regions must be constituted as an effect of analysis; they are thus defined in relation to spatial uneven development in the process of accumulation and its effects on social (including political) relations' (Massey, 1978: 110; see also Massey, 1984: 123, 298–300). As this quotation attests, Massey's commitment to understanding regions in terms of relations rather pre-dates the *fin de siècle* enthusiasm of many geographers for the idea of relational space. It drew on and pushed further earlier attempts to think about the contested geographies of production and labour practices (Beynon, 1973, Massey, 1995b: 86–88). And, as Massey herself notes somewhat wistfully in the opening chapter of the second edition of *Spatial Divisions of Labour*, 'to say that social space is relational has become commonplace. But things are more easily said than fully understood or thought through into practice' (Massey, 1995: 1).

Two essays in this volume respond to the challenge of thinking through the implications of a relational understanding of space. In her essay 'The Political Challenge of Relational Territory', Elena dell'Agnese argues that thinking relationally about space need not spell the end of the concept of territory in the sense of a bounded portion of space, provided the space in question is understood in relational terms. Moreover the converse is also true – the apparent 'hardening' of territorial boundaries in recent years in response to heightened fears about security does not disprove relational theories of space. 'Territorial' and 'relational' understandings of space are not necessarily mutually incompatible.

In his contribution, Allan Cochrane notes that Massey's 'conclusions may seem rather uncontroversial, saying no more than that it is necessary to understand that there are no geographical absolutes since space is necessarily defined in the relations between actors, objects and context'. This apparently simplicity is deceptive, however, and 'once one begins to take these points seriously then it is the ways in which these relations work out that matter as much as the patterns that emerge'. Cochrane takes inspiration from Massey's emphasis on multiplicity – the coming together in place of diverse social and economic trajectories – to argue that regions can usefully be understood as 'assemblages'. Assemblage thinking is increasingly prevalent in geography (Anderson and McFarlane, 2011; McFarlane, 2011) and usually takes its cue from two sources: the actor–network theory of Bruno Latour (Latour, 2005) and/or the notion of 'machinic assemblage' developed in the work of Gilles Deleuze and Félix Guattari (e.g., Deleuze and Guattari, 1987). There are also, though, affinities between the notion of assemblage and Althusser's anti-essentialist concept of overdetermination which inspired Massey. In Cochrane's hands the idea of assemblage is used as a way of building on Massey's work to understand the contingent and provisional ways in which regions are formed and re-formed.

The significance of contingency and specificity was hotly disputed in geography and urban and regional studies in the decade after the publication of *Spatial Divisions of Labour*. These arguments and a series of related debates were promoted by an upsurge in research into the effects of economic restructuring on localities in Britain and the role of geographical difference in shaping economic change. Several such 'locality studies' were undertaken as part of a research programme on the Changing Urban and Regional System funded by the UK's Economic and Social Research Council in the second half of the 1980s. The programme was initiated by Doreen Massey, who drew up the original proposal (Massey, 1991a), though the studies themselves were undertaken by a number of other research teams. The research was criticised by some on the left for overemphasising the local and the empirical at the expense of wider processes and general theory. In a 1991 commentary on the 'political place of locality studies' Massey voiced criticisms of *both* the neglect of the political in some localities research *and* the view of its critics that a focus on the local and the empirical was necessarily anti-progressive. (Massey also emphasised that the local, the specific and the empirical are not the same things – it is possible to study the global scale empirically and to do theoretical work on the local.)

At stake here was also the critical debate prompted by *Spatial Divisions of Labour*, most notably David Harvey's critique of Massey's attention to contingency and specificity. Harvey argued that in the text 'all theorising disappears between a mass of contingent labour–management relations in place' (Harvey, 1987: 369). These debates speak to broader struggles over the terms of political-economic geographical analysis. Julie Graham has argued that *Spatial Divisions of Labour* prefigured challenges to capital-centred accounts of left geography (Graham, 1998). Such challenges were to become central to debates in critical geographical work through the 1990s and 2000s, particularly through labour geography (Herod, 2001). By decentring capital, as Jamie Peck notes here, Massey sought to open up 'a different kind of politics of possibility' where 'new terrains of political potential are always under construction'.

Her refusal to see the local and empirical as secondary to supposedly more powerful and theoretically generative forces drew on an enduring suspicion of the counter-position of local and global, place and space. Indeed a key characteristic of Massey's work throughout her career has been her concern for the empirical – even in her most abstract and philosophical writings she is at pains to exemplify her arguments with reference to current political events, autobiographical experience and geographical case studies. There is never a sense in *Spatial Divisions of Labour* or elsewhere that labour is purely a theoretical category. Rather, Massey cares passionately – and politically – about the people whose lives (working and otherwise) animate the geographies of division and connection that are at the heart of her academic concerns. Class inequality (never just 'deprivation' or 'poverty') is central to her understanding of social and economic geography. As Jamie Peck notes in his chapter here, *Spatial Divisions of Labour* can be read in part as having provided the foundations for, and an early contribution to, what has since become known as Labour Geography. Moreover, Peck suggests that recent reassessments of the project of 'capital-L' Labour Geography would benefit from a sharper recognition of these antecedents, and from a greater attention to the 'evolving strategies of capital and the state' that Massey has done so much to unpack.

Questions of regional inequality and the class and gender divisions of labour markets have remained important to Massey throughout her career (e.g., Allen *et al.*, 1998; Amin, Massey and Thrift, 2003; Massey, 2007). In *World City* (Massey, 2007) she examines the implications of London's spectacular growth for other parts of the UK (and other parts of the world) as well as the exacerbation of inequalities along class, gender and ethnic lines within the city. Whether we choose to see London as a territory, a region, a labour market or an assemblage, *World City* demonstrates the power of Massey's relational thinking, drawing attention to London's connections with and dependencies on other places – some near and some far away.

Her engagements with regional inequality continue to have impacts on political debate. The contention in the pamphlet *Decentering the Nation* on regional inequality, co-written with Ash Amin and Nigel Thrift (2003), that Britain has an 'undeclared' 'London/ South-East Regional Policy' was picked up in *Clear Red Water* by Nick Davies and Darren Williams (2009: 112). This book assesses the attempt by Rhodri Morgan's devolved Labour administration in Wales to forge political alternatives to New Labour, particularly through a strong commitment to 'progressive univer- salism'. Davies and Williams drew on *Decentering the Nation* to challenge hegemonic political narratives that 'the efficient South-East is subsidising Wales, Scotland and the North of England' (Davies and Williams, 2009: 112). This opens up an important challenge to dominant ways of configuring the contemporary regional geographies of the UK.

Reconceptualising Place

In 'A Global Sense of Place', one of her best known and most influential essays, Massey argues that it would be impossible to understand Kilburn, the neighbourhood in north-west London where she lives, 'without bringing into play half the world and a considerable amount of British imperialist history' (Massey, 1994a: 154). Her vivid account of walking down the Kilburn High Road, talking to her newsagent during the first Gulf War – 'silently chafing at having to sell the *Sun*' – challenged pervasive accounts of the erosion of place in 'global times'. Such narratives about the loss of distinctive places continue to structure debates about globalisation. Massey begins by questioning the tendency to counterpose the local and the global and idealised notions of place with the sense of fragmentation and disruption associated with time–space compression. She combines an insistence on the persistence of place as significant in global times with a refusal to subscribe to 'static and defensive' notions of 'place' (Massey, 1994a: 153).

Here the concept of power-geometry makes an early appearance, referring to the ways in which different social groups are situated differently in relation to global flows and connections. The themes of difference, inequality, international connections and spatial thinking come together in an argument that Massey went on to develop and refine in a number of subsequent writings, including *World City* and *For Space*. Massey writes about how the place that is Kilburn is complex and heterogeneous, and intimately connected to other places and to the flows that link them together. This essay is the most influential of a number of her interventions which have sought to reconceptualise place (see also Massey, 1993a, 1994b). As Rogério Haesbaert argues here, the idea of a 'global sense of place' has the status of a 'universally recognised contribution to the geographic debate ... even if it is to argue with the idea' (for critical discussion see McGuinness, 2000; Dirlik, 2001; Hardt and Negri, 2001).

In her contribution Jane Wills reflects on Massey's understanding of the politics of place. Wills contends that, notwithstanding the significance of extra-local and international flows and connections in constituting places, it is important not to lose sight of the role of locally rooted traditions that may provide valuable political and democratic resources. She emphasises 'the importance of place in the formation of social relationships, the creation of the public realm and the vitality of political life'. Wills suggests that this stems from the way that place can provide a forum for our encounters with – and negotiations across – difference, but also from the role that place plays in allowing collective mobilisation around common interests.

The idea of a global sense of place is taken up explicitly by Rogério Haesbaert. As Haesbaert notes, the meaning of the English word 'place' is often carried in Latin languages by the words for territory (territorio, território, territoire etc.). This can lead to confusion in translation since the English term 'territory' carries a much stronger sense of a coherent, bounded space than its apparent equivalents in Portuguese, Spanish, French and Italian. Haesbaert proposes a concept of territory that, like Massey's notion of place, is processual, porous, and endowed with multiple identities. He proposes the idea of 'multi-territoriality', which may be 'simultaneous', involving the experience of (or interaction with) diverse territories/places at the same time without needing to move between them, or 'successive' involving movement among multiple networked territories/places. Like Wills, Haesbaert stresses the progressive potential of 'relative territorial closure' in the context of wider struggles and networks of 'transterritorial' connections.

Massey's engagements with place have been suspicious of those on the left who romanticise or seek to construct authentic accounts of place. In her essay 'Travelling Thoughts', for example, she contests Raymond Williams's telluric and male-centred account of place in his novel *Border Country* arguing that 'Places change; they go on without you. Just as Mother has a life of her own' (Massey, 2000a: 230, see also Massey, Quintas and Wield, 1991). The importance of her engagements with place for feminist politics and practice are one of many key themes that Wendy Harcourt, Alice Brooke Wilson, Arturo Escobar and Dianne Rocheleau discuss in a multi-voiced series of reflections. They explore Massey's influence on their work on the agency of women in realising the progressive potential of place and on the local food movement. Harcourt describes how Massey encouraged them not to counterpose space and place and not to see place and the local as necessarily progressive in terms of gender relations. Similarly, Wilson highlights Massey's warning not to romanticise the local or indulge in an 'a priori politics of topographies' (Massey, 2005: 172). For Escobar, Massey's 'feminist politics of place showed that, without shunning political economy, it is possible to find profound meaning and hope in the struggles carried out by peoples in places'. He takes three key lessons from Massey's work: the need to think of place within wider networks of relations; that places are always sites of negotiation and transformation; and that a relational understanding of place 'ineluctably calls for a politics of responsibility'. Rocheleau's notion of 'rooted networks' seeks to capture some of these insights. Her work on feminist political ecology emphasises the importance of the biophysical dimensions of the relational connections through which places are formed and transformed.

Indeed a key way in which Doreen Massey has developed her theorising about place is through moving beyond the 'realm of human social relations' (Massey, 2006: 34). The biophysical world is of great importance to her and, as her writings attest, that importance is simultaneously personal, conceptual and political. As human and physical geography have grown apart methodologically and institutionally, Massey has sought to promote dialogue and debate between them, and to question the nature/culture binary in her own work. Steve Hinchliffe's chapter also takes its cue from Massey's essay 'A global sense of place', but turns it to reflect on her understanding of non-human nature and the physical materiality of the world. Hinchliffe describes how Massey extends her non-foundational understanding of place beyond culturally diverse urban areas such as Kilburn to encompass even such apparently fixed and timeless physical landscapes as the English Lake District.

If place for Massey is the product of multiple heterogeneous associations and relations stretching beyond the here and how, then (bio-)physical materials and processes are integral parts of that heterogeneity and those relations, provided they are understood in terms of the appropriate temporal and spatial scales. In geological time, the seemingly ancient and permanent rocks and mountains of the Lake District are young and have come from elsewhere, to say nothing of the Icelandic landscapes in Olafur Eliasson's haunting photographs. While few natural scientists will quarrel with Massey's characterisation of the physical world as dynamic and historical, some may baulk at Massey's questioning of the foundational character of scientific knowledge itself. Nevertheless, as Hinchliffe puts it, 'non-foundational approaches to nature and knowledge necessitate a serious engagement with the world [that is] anything but "grounded" in the sense of fixed or stable, but it is "grounded" in the sense of dealing with the complex multiplicities involved in the making of realities'. For Massey, this unsettles critiques of 'a global sense of place' which contended that an engagement with 'nature' would restore a lost groundedness to her account of place (Massey, 2006: 34). This is part of an attempt to construct conversations across human and physical divides in geography (Massey, 1999b).

That she chose to reimagine landscape through discussion of Skiddaw, 'the grey slab of a mountain' outside the town of Keswick in the English Lakes, reflects Massey's roots in England's North-West. She was brought up on the Wythenshawe estate in Manchester. The creation of this estate was, as Massey notes, the result of a struggle between 'a collection of social-ists and progressives battling to win more, and healthy, space for the city's working class' and the 'local people of rural north Cheshire' including large landowners (Massey cited by Watson, 1999: 212). The socialist feminist Hannah Mitchell, a radical suffragette and councillor for the Independent Labour Party, described the purchase of the Wythenshawe estate as one of the two outstanding events of her council life (Mitchell, 1977: 220). Massey notes in the conversation published here how this background has shaped her political outlook. She describes having 'lived with, through and kind of in combat with regional inequality since my childhood'.

Political Trajectories

For Massey, a strict division between political and intellectual work has never made much sense. Her intellectual work and projects have been directly animated by political engagements and problematics, and vice versa. She argues that some of the 'most intense theoretical and empirical debates we had were motivated by politics "outside"' the academy (Massey et al., 2009: 407). Her pedagogy has functioned in similar terms, both in terms of its style, particularly through the collective practices of course pro-duction at the Open University, and through its content. The Social Science Foundation Course D102, which Doreen played a significant part in, was to cause the Thatcherite ideologue Sir Keith Joseph 'sleepless nights' (Thompson, 1987: 129).

Her first book, *Capital and Land*, co-written with Alejandrina Catalano, was prompted by a set of questions around the politics of landownership in Britain. It explored the implications of the 'entry into direct landownership of agencies such as property companies, insurance companies and pension funds'. These questions remain prescient. The intellectual rationale for the book was that 'the analysis of this phenomenon 'on the left' appeared to be confused' (Massey and Catalano, 1978: 1). In *For Space* she avers that many of her engagements with spatial thinking have come out of 'Picking away at things that don't seem quite right. Losing political arguments because the terms don't fit what it is you're struggling to say' (Massey, 2005: 1).

Sophie Bond and Sara Kindon in their chapter argue that 'openness and struggle' are central to Massey's ways of thinking and acting the political. Her feminism has shaped a particular ethos of political engagement, and a questioning of hierarchical left practices. This has been defined by a commitment to a particular style of democratic left political practice. Bond and Kindon develop a sense of what this opens up politically through a discussion of important political tensions in post-colonial Aotearoa New Zealand. They demonstrate how a concern with dynamic and multiple political trajectories offers new possibilities for thinking about the coexistence of different groups within contemporary Aotearoa New Zealand.

A stress on the openness of the political is also central to the contribution of Ash Amin and Nigel Thrift, who observe that Massey's 'political stance is principled but not orthodox, coherent but not dogmatic, critical but hopeful'. Amin and Thrift argue that her work 'has displayed a play between constancy and innovation, critique and proposition, and particularity and more general tracing of political contours that would serve the Left well in reaffirming its future necessity and relevance'. They are particularly concerned to think through the implication of left politics in a 'world in which human projects can only ever unfold in uncertain and open-ended ways'.

Massey's concern with the happenstance and generative practices through which political trajectories are linked has been a key way in which she has contributed to a rethinking of left political practices. Her contributions have been defined by openness to different ways of envisioning the left; openness to diverse constituencies and modalities of doing politics; and crucially, a geographical openness. Massey's political engagements have been structured by various forms of internationalist commitment and, as the last section emphasised, to viewing place-based politics in open rather than bounded ways. This expansive imagining of left politics was integral to the founding of *Soundings: A Journal of Politics and Culture* in 1995, along with her co-editors Stuart Hall and Michael Rustin.

The positioning statement of the journal, 'Uncomfortable Times', noted that 'change can be achieved in many social spaces besides that which is normally designated as political' and noted a commitment to 'report specific interventions in what is often described as "civil society", and support them in the cause of a fuller process of cultural and social democratisation' (Hall, Massey and Rustin, 1995: 15). This approach contrasted markedly with the political practices and approach of other prominent left journals, most notably *New Left Review*, grouped around its 'steely Leninist core' (Thompson, 2007: 156).[3] Most recently, *Soundings* has been at the forefront of attempts to construct a conjunctural analysis of the economic crisis and to dissect the reworking of neoliberalism being produced by the UK Coalition government (see Massey, 2010; Hall, 2011).

Massey's commitment to such openness, and to the productive articulation of diverse political trajectories, has been shaped as much by her engagements with political activity as by abstract theoretical principles. This is perhaps why for Massey a sense of the constructed and contingent character of political identities and alliances has been something which she has embraced as having political potential rather than as diluting or threatening fixed commitments and positions. Such political inventiveness, for example, shaped Massey's involvement in the Greater London Council, a key part of the 'new urban left'. Ken Livingstone here recalls making her an early appointment to the Greater London Enterprise Board. Hilary Wainwright's contribution notes the importance of anti-racism and anti-sexism to the political cultures of this 'new urban left' (see also Peck, 2011: 44–51).

During the 1984–1985 miners' strike Massey and Wainwright were both involved in the network of support groups that sprang up in solidarity with the struggle of the National Union of Mineworkers against pit closures. The year-long dispute came to be defined as a broader struggle against the devastating impact of Thatcherite monetarism. Massey and Wainwright's co-written chapter in the collection *Digging Deeper*, one of the first attempts to grapple with the broader implications of the strike, indicates Massey's engagement with the generative character of political activity. They argued that the support groups emphasised how 'industrial action' and 'new social movements' could be mutually dependent rather than antagonistic (Massey and Wainwright, 1985: 168).

Acknowledging that 'the existing institutions of labour are old-fashioned and sectional', they argued that 'what the miners' strike has shown is that these institutions can be superseded and challenged without abandoning class politics' (Massey and Wainwright, 1985: 168). This positions solidarity as a practice of articulation where different elements are reworked and redefined through political struggle. Hywel Francis, the Welsh labour historian and activist who shared platforms with Doreen during the strike, writes of the emergence of 'a broad democratic alliance of a new kind – an anti-Thatcher alliance – in which the organised working class had a central role but a role which it would have to earn and not assume' (Francis, 2009: 62). These interventions offered different possible futures for Labour politics than the 'near obsessional bleaching of class' which Wainwright argues came to define 'New Labour'.

Massey and Wainwright note how these solidarities partly reconfigured the terms of place-based politics, especially in terms of gender relations and ethnicity (see also Massey and McDowell, 1995). Massey's work has never been defined by an abstracted, placeless politics of solidarity or internationalism. Rather, her commitment to a 'networked, practiced internationalism' has challenged 'the dominant geographical imaginary which understands the world in terms of scales and nested hierarchies' and situates internationalism as necessarily locally constituted (Massey, 2007: 184). This opens up diverse agency in shaping internationalist politics and decentres the importance of left elites in shaping internationalism (Massey, 2006).

Massey's approach asserts the diverse forms of agency and solidarity shaped through internationalist practices and political cultures (see also Featherstone, 2012).

The importance of Massey's networked account of internationalism for engagements with the spatial politics of global justice networks is discussed here by Andy Cumbers and Paul Routledge. They argue that, 'importantly, and unlike many others who write about globalisation, her accounts of extended spatialities are nuanced by a continuing awareness of the territorial politics of place and how this in terms produces globalisation'. Massey's argument that globalisation was produced through situated local practices opens up possibilities for ways of thinking the politics of how globalisations might be shaped in different ways. Cumbers and Routledge discuss how different activist networks sought to construct such alternative globalisations, usefully pointing to some of the tensions involved in such projects.

A key aspect of Massey's internationalist political work has been a longstanding engagement with political struggles and movements in Central and Latin America. In the mid-1980s Doreen spent a year in Nicaragua, which included time spent working at the Instituto Nacional de Investigaciones Económicas y Sociales in Managua, where she collaborated with Marielos de los Angeles and Ixy Martínez. Her involvement was part of the concerted internationalist solidarity movements active in support of Central American revolutionaries in El Salvador, Nicaragua and Guatemala. Her writings on the Sandinistas' political engagements and territorial politics recognised both the forms of 'self-confidence and sense of rights' which were 'won through the revolutionary process' and the dilemmas posed by forging such a process in the 'shadow' of the US-sponsored Contras (Massey, 1986: 328; 1987).

More recently, her ideas have had an important influence on the Bolivarian Revolution in Venezuela. Hugo Chávez adopted the term 'power-geometry' as one of the drivers of the Venezuelan revolutionary process. In 2007 five motors of the revolution were identified, Motor-4 being 'we have to build a new power-geometry'. She suggests here that the concept was transformed by being used as part of a directly political process. The implications of this attempt to generate new forms and geographies of political participation are taken up in depth here by Ricardo Menéndez, the minister for science and technology in the Venezuelan government and a geographer by training. Asserting the importance of 'geography as a component of revolutionary strategy', he probes questions about the relations between geography, participation and the formation of popular state power.

Massey's engagements with internationalist political activity have emphasised the importance of ongoing forms of connectedness. This disrupts constructions of internationalism as being about the exoticised struggles of 'distant others'. One key way in which she explores such forms of connectedness is through what she terms the 'politics of place beyond place'. Hilary Wainwright explores this aspect of Doreen Massey's work, arguing that it provides 'intellectual tools for understanding the contested nature of dependencies, inter-connections and effects between the local and the global'. Massey has sought to focus on a politics of the different forms of connections that shape alternative ways of doing globalisation. She drew attention, for example, to the importance of the London– Caracas deal negotiated by Ken Livingstone and Hugo Chávez. This link challenged neoliberal conventions through 'a barter arrangement in which Venezuela would send cheap oil to London in return for advice and experience in the areas of transport planning, housing, crime, waste-disposal, air quality and adult education' (Massey, 2007: 199). She argued that a key aspect of the deal was its 'jolt to the dominant, imperial, geographical imagination' and 'the implicit question of the identity of places that went along with that' (Massey, 2011b: 12).

It is appropriate for Doreen to have the last word in the collection. The book closes with 'Stories So Far', a wide-ranging conversation with Doreen which provides a multi-faceted account of her work, ideas and political engagements. The interview started life when she was presented with an honorary doctorate by Glasgow University. A conversation between geographers from Glasgow's Human Geography Research Group and Doreen took place, which was later published in the *Scottish Geographical Journal* (Massey *et al.*, 2009). 'Stories So Far' combines selected material from this discussion with a conversation which took place between the editors and Doreen in London in January, 2011.[4] The interview engages with some of the key trajectories that have run through Doreen Massey's work and also explores her recent projects, such as her involvement in the making of the film *Robinson in Ruins* with the film-maker Patrick Keiller.

We are keenly aware that this book is one of just many possible takes on Doreen Massey's work. In this sense the book is provisional. It is one of a number of attempts to engage with her work that includes a recently published book by Spanish geographers (see Albet and Benach, 2012). *Spatial Politics* seeks to open up engagements, rather than to generate a closed or final perspective on Doreen Massey's many interventions. In the spirit of Massey's political and theoretical engagements the book seeks to be neither introspective nor retrospective. Rather, the collection seeks to advance in a critical – but sympathetic – vein Doreen Massey's intellectual, theoretical and political agendas. As Jamie Peck argues, this 'need not be an exercise in intellectual or political nostalgia'. Serious reflection on Doreen Massey's work instead offers important resources and intellectual tools for reshaping our geographical and political futures.

Notes

1   This account draws on the discussion of Massey's talk in Miller (2011). For Massey's reflections on the politics of the crisis, see Massey (2010, 2011a).

2   Her intellectual positions have frequently been developed through collaborative work with colleagues at the Open University, most notably a long and productive association with John Allen.

3   Doreen Massey joined the editorial collective of *New Left Review* in 1990 as part of what Patrick Wright described as a 'genuine attempt' to reach beyond the core (Wright, 1993). She was one of the 13 members of the editorial board, including her future *Soundings* co-editor Mike Rustin, who resigned en bloc in 1993 during disputes over the direction/funding of the journal (Thompson, 2007: 157). Arguably the character of *Soundings* owed much to *Marxism Today*, but also to the style of *Universities and Left Review*, edited by Stuart Hall, which merged with the *New Reasoner* edited by John Saville and E.P. Thompson to become the *New Left Review* in 1960.

4   Many thanks to Mo Hume, Danny Mackinnon, Geraldine Perriam, Chris Philo and Jo Sharp for permission to reproduce the questions they contributed to this original interview.

Part One
Space, Politics and Radical Democracy

Chapter One
Space, Hegemony and Radical Critique
Chantal Mouffe

Introduction

What kind of radical critique is still possible in our post-political world where we are constantly told that there is no alternative to the current mode of neoliberal globalisation? To approach this question we first need to ask how critique should be envisaged. Indeed there are many different understandings of the nature of critique, and the forms that correspond to them are very diverse. Should we envisage the activity of critique in terms of judgement or in terms of practice? Is it, as it is often claimed, a self-conscious activity linked to the Enlightenment and characteristic of modernity? Besides, as Foucault has rightly noted, critique cannot be defined apart from its objects and is therefore condemned to dispersion. Centring my investigation on social criticism will limit the field of possible meanings, but crucial disagreements will nonetheless remain, for instance between Habermas, who argues that social criticism depends on a form of critical theory of society – of the type of his theory of communicative action – providing the ground for making strong normative judgements, and others who, like Foucault, see criticism as a practice of resistance.

However, I will approach this issue in a different way. Since my aim is to scrutinise the relation between social criticism and radical politics, I have chosen to examine one of the currently most fashionable views of social criticism today, which visualises radical politics in terms of desertion and

Spatial Politics: Essays for Doreen Massey, First Edition.
Edited by David Featherstone and Joe Painter.

exodus, contrasting it with the hegemonic approach that I have been advocating in my work. I intend to bring to the fore the main differences between those approaches, which could roughly be distinguished as 'critique as withdrawal from' and 'critique as engagement with', and to show how they stem from conflicting theoretical frameworks and understandings of the political. I argue that the form of radical politics advocated by post-operaist thinkers like Negri and Virno is informed by a flawed understanding of the nature of space (see Massey, 2005: 174–175).[1] I contend that this is an understanding that does not allow them to acknowledge the ineradicable dimension of antagonism and to grasp the dimension of 'the political'.

The work of Doreen Massey has been very important to me for thinking about politics because she really made me realise the importance of the spatial dimension in politics, something I had not been aware of previously. There are two implications of this position that I find to be particularly significant for thinking about democratic politics. First, Massey has insisted that space is a dimension of multiplicity. She has always insisted that space and multiplicity are co-constitutive. Space poses the question of how we are going to live together. This is a crucial question, of course, for democratic politics. Second is the idea that space is the product of relations and practices and that we need to acknowledge our co-constitutive interrelatedness, and that implies a spatiality. I think this is extremely important. So, if we are going to think about how we are going to live together there is necessarily a dimension of spatiality; this is something that many political theorists do not fully realise. In this chapter I use these insights to engage with the current state of radical politics.

Critique as Withdrawal From

The model of social criticism and radical politics put forward by Michel Hardt and Antonio Negri in their books *Empire*, *Multitude* and *Commonwealth* calls for a total break with modernity and the elaboration of a postmodern approach. Such a break, they say, is required because of the crucial transformations undergone by our societies since the last decades of the twentieth century. According to them, those changes, which are the consequences of the process of globalisation and transformations in the work process brought about by workers' struggles, can be broadly summarised in the following way:

1   Sovereignty has taken a new form composed of a series of national and supranational organisms united under a single logic of rule. This new global form of sovereignty, which Hardt and Negri call 'Empire', has replaced the stage of imperialism, which was still based on the attempt by nation states to extend their sovereignty beyond their borders. In

contrast to what happened in the stage of imperialism, the current Empire has no territorial centre of power and no fixed boundaries; it is a decentered and deterritorialised apparatus of rule that progressively incorporates the entire global realm with open, expanding frontiers.

2   This transformation corresponds to the transformation of the capitalist mode of production in which the role of industrial factory labour has been reduced and priority given to communicative, cooperative and affective labour. In the postmodernisation of the global economy, the creation of wealth tends towards biopolitical production. The object of the rule of empire is social life in its entirety; it presents the paradigmatic form of biopower.

3   We are witnessing the passage from a 'disciplinary society' (Foucault) to a 'society of control' (Deleuze) characterised by a new paradigm of power. In the disciplinary society, which corresponds to the first phase of capitalist accumulation, command is constructed through a diffuse network of dispositifs or apparatus that produce and regulate customs, habits and productive practices, with the help of disciplinary institutions like prisons, factories, asylums, hospitals, schools and so forth. The society of control, in contrast, is a society in which mechanisms of command become immanent to the social field, distributed to the brains and bodies of the citizens. The modes of social integration and exclusion are increasingly interiorised through mechanisms that directly organise the brains and bodies. This new paradigm of power is biopolitical in nature. What is directly at stake in power is the production and reproduction of life itself.

4   Hardt and Negri assert that the notions of 'mass intellectuality', 'immaterial labour' and 'general intellect' help us to grasp the relation between social production and biopower. The central role previously occupied by the labour power of mass factory workers in the production of surplus value is today increasingly filled by intellectual, immaterial and communicative labour power. The figure of immaterial labour involved in communication, cooperation and the reproduction of affects occupies an increasingly central position in the schema of capitalist production.

5   In the passage to postmodernity and biopolitical production labour power has become increasingly collective and social and a new term is needed to refer to this collective worker: the 'Multitude'. The Multitude, say Hardt and Negri, called Empire into being and they present the construction of Empire as a response to the various machines of power and the struggles of the Multitude. They claim that the passage to Empire opens new possibilities for the liberation of the Multitude. According to them, globalisation, in so far as it operates a real deterritorialisation of the previous structures of exploitation and control, is a condition of the liberation of the Multitude. The creative forces of the Multitude that sustain Empire are capable of constructing a counter-empire,

an alternative political organisation of the global flows of exchange and globalisation, so as to reorganise them and direct them towards new ends.

At this point it is worth introducing the work of Paolo Virno to complement the picture. Virno's analyses in his book *Grammar of the Multitude* dovetail in many respects with those of Hardt and Negri, but there are also some significant differences. For instance, he is much less sanguine about the future. While Hardt and Negri have a messianic vision of the role of the Multitude, which will necessarily bring down Empire and establish an 'Absolute Democracy', Virno sees current developments as an ambivalent phenomenon and he acknowledges the new forms of subjection and precarisation which are typical of the post-Fordist stage. It is true that people are not as passive as before, but it is because they have now become active actors of their own precarisation. So, instead of seeing the generalisation of immaterial labour as a type of 'spontaneous communism' like Hardt and Negri, Virno sees post-Fordism as a manifestation of the 'communism of capital'. He notes that, today, capitalistic initiatives orchestrate for their own benefits precisely those material and cultural conditions which could, in other conditions, have opened the way for a potential communist future.

When it comes to envisaging how the Multitude could liberate itself, Virno declares that the post-Fordist era requires the creation of a 'Republic of the Multitude' – by which he understands a sphere of common affairs which is no longer state run. He proposes two key terms to grasp the type of political action characteristic of the Multitude: exodus and civil disobedience. Exodus is, according to him, a fully fledged model of political action, capable of confronting the challenges of modern politics. It consists in a mass defection from the state aiming at developing the public character of Intellect outside of work and in opposition to it. This requires the development of a non-state public sphere and a radically new type of democracy, framed in terms of the construction and experimentation of forms of non-representative and extra-parliamentary democracy organised around leagues, councils and soviets. The democracy of the Multitude expresses itself in an ensemble of acting minorities which never aspire to transform themselves into a majority and develop a power that refuses to become government. Its mode of being is 'acting in concert' and while tending to dismantle the supreme power, it is not inclined to become state in its turn. This is why civil disobedience needs to be emancipated from the liberal tradition within which framework it is generally located. In the case of the Multitude it does not mean any more ignoring a specific law because it does not conform to the principles of the constitution. This would still be a way of expressing loyalty to the state. What is at stake is a radical disobedience which puts in question the state's very faculty of command.

With respect to the question of how to conceive the democracy of the multitude, there is a basic agreement among Hardt/Negri and Virno. In both cases we find a rejection of the model of representative democracy and the drawing of a stark opposition between the Multitude and the People. For them, the problem with the notion of the people is that it is represented in a unity, with one will, and that it is linked to the existence of the state. The Multitude, on the contrary, shuns political unity. It is not representable because it is a singular multiplicity. It is an active self-organising agent that can never achieve the status of a juridical personage and can never converge in a general will. It is anti-state and anti-popular. They state that the democracy of the Multitude cannot be visualised in terms of a sovereign authority that is representative of the people, and they call for new forms of democracy which are non-representative. Since in Empire there is no more outside, the struggles must be against in every place. This 'being against' is for them the key to every political position in the world and the Multitude must recognise imperial sovereignty as the enemy and discover adequate means of subverting its power. Whereas in the disciplinary era sabotage was the fundamental form of resistance, they claim that in the era of imperial control it should be desertion. It is indeed through desertion, through the evacuation of the places of power, that they think that battles against Empire might be won. Desertion and exodus are for them a powerful form of class struggle against imperial postmodernity.

As we can see, according to this model, the activity of critique corresponds to a form of negation which consists in withdrawal from existing institutions. At the core of the agreement among the theorists of the Multitude we find a celebration of the process of 'deterritorialisation' which is presented as providing the conditions for the disappearance of states and the emergence of an increasingly 'smooth' democratic world beyond sovereignty and the constraints of state power. It is for this reason that they want the Multitude to liberate itself from all forms of belonging and that they denounce local and regional attachments as obstacles to the globalised absolute democracy that they are advocating.

In my view, one of the main problems with this approach comes from the inadequate conception of spatiality that informs their view of globalisation, which they claim is leading to the establishment of a 'smooth' space. This idea needs to be challenged because it has direct consequences for their mistaken conception of politics. To criticise such a flawed approach, the work of Doreen Massey is of great relevance. By bringing to the fore the fact that space is always striated because it is a product of relations and struggles, her concept of 'geometries of power' highlights the way in which power plays a central role in the construction of spatialised social practices (Massey, 1993a). She argues that we need a local politics that thinks beyond the local, acknowledging that the local is globally produced and the global locally produced (Massey, 1991b, 2007). This helps us to grasp the importance of the spatial dimension in politics and to envisage a politics of place aiming at both defending and challenging the nature of the local. Contrary to those who think only in terms of the global and dismiss local and regional attachments, Massey's approach allows us to scrutinise the role of the local in the construction of wider power geometries, thereby opening new avenues for political engagement and challenging the strategy of exodus and desertion advocated by authors like Hardt, Negri and Virno.

Critique as Hegemonic Engagement With

Before presenting my own view about the form of social criticism best suited to radical politics today, I would like to state that I recognise the necessity of taking account of the crucial transformations in the mode of regulation of capitalism brought about by the transition from Fordism to post-Fordism. I agree with the importance of not seeing those transformations as the mere consequence of technological progress and of emphasising their political dimension. I consider, however, that the dynamics of this transition are better apprehended within the framework of the theory of hegemony that we have put forward in *Hegemony and Socialist Strategy: Towards a Radical Democratic Politics* (2001 [1985]), written jointly with Ernesto Laclau. What I want to stress is that many factors have contributed to this transition and that it is necessary to recognise its complex nature. My problem with the operaist and post-operaist view is that, by putting an almost exclusive accent on the workers' struggles, they tend to see this transition as if it was driven by one single logic: workers' resistance to the process of exploitation forcing the capitalists to reorganise the process of production and to move to post-Fordism with its centrality of immaterial labour. In their view, capitalism can only be reactive and they refuse to accept the creative role played both by capital and by labour. What they deny is, in fact, the role played in this transition by the hegemonic struggle.

According to the approach that I am advocating, the two key concepts for addressing the question of the political are 'antagonism' and 'hegemony'. On one side, it is necessary to acknowledge the dimension of the political as the ever-present possibility of antagonism and this requires, on the other side, coming to terms with the lack of a final ground and the undecidability that pervades every order. This means recognising the hegemonic nature of every kind of social order and envisaging society as the product of a series of practices whose aim is to establish order in a context of contingency. The practices of articulation, through which a given order is created and the meaning of social institutions fixed, we call 'hegemonic practices'. Every order is the temporary and precarious articulation of contingent practices. Things could always have been otherwise and every order is predicated on the exclusion of other possibilities. It is always the expression of a particular structure of power relations. What is at a given moment accepted as the 'natural order', jointly with the common sense that accompanies it, is the result of sedimented hegemonic practices; it is never the manifestation of a deeper objectivity exterior to the practices that bring it into being. Every hegemonic order is susceptible of being challenged by counter-hegemonic practices which attempt to disarticulate it in order to install another form of hegemony.

I submit that it is necessary to introduce this hegemonic dimension when one envisages the transition from Fordism to post-Fordism. This means abandoning the view that one single logic (workers' struggles) is at work in the evolution of the labour process, and acknowledging the proactive role played by capital. In order to do this we can find interesting insights in the work of Luc Boltanski and Eve Chiapello, who, in their book *The New Spirit of Capitalism*, bring to light the way in which capitalists manage to use the demands for autonomy of the new movements that developed in the 1960s, harnessing them in the development of the post-Fordist networked economy and transforming them into new forms of control. What they call 'artistic critique', to refer to the aesthetic strategies of the counter-culture (the search for authenticity, the ideal of self-management, the anti-hierarchical exigency), were used to promote the conditions required by the new mode of capitalist regulation, replacing the disciplinary framework characteristic of the Fordist period.

From my point of view, what is interesting in this approach is that it shows how an important dimension of the transition from Fordism to post-Fordism is the process of discursive rearticulation of existing discourses and practices, allowing us to visualise this transition in terms of a hegemonic intervention. To be sure, Boltanski and Chiapello never use this vocabulary, but their analysis is a clear example of what Gramsci called 'hegemony through neutralisation' or 'passive revolution' to refer to a situation where demands which challenge the hegemonic order are recuperated by the existing system by satisfying them in a way that neutralises their subversive potential. When we apprehend the transition from Fordism to post-Fordism within such a framework, we can understand it as a hegemonic move by capital to re-establish its leading role and restore its challenged legitimacy.

It is clear that, once we envisage social reality in terms of hegemonic practices, the process of social critique characteristic of radical politics cannot consist any longer in a withdrawal from the existing institutions, but in an engagement with them in order to disarticulate the existing discourses and practices through which the current hegemony is established and reproduced, with the aim of constructing a different one. Such a process, I want to stress, cannot merely consist in separating the different elements whose discursive articulation is at the origin of those practices and institutions, or for that matter in 'deserting' them. The second moment, the moment of rearticulation, is crucial. Otherwise we will be faced with a chaotic situation of pure dissemination, leaving the door open for attempts at rearticulation by non-progressive forces. Indeed, we have many historical examples of situations in which the crisis of the dominant order led to right-wing solutions. It is therefore important that the moment of 'de-identification' be accompanied by a moment of 're-identification', and that the critique and disarticulation of the existing hegemony goes hand in hand with a process of rearticulation. This is something that is missed by all approaches in terms of reification or false consciousness which believe that it is enough to lift the weight of the dominant ideology in order to bring about a new order, free from oppression and power. It is also missed, albeit in a different way, by the theorists of the Multitude who believe that its oppositional consciousness does not require political articulation. For the hegemonic approach, social reality is discursively constructed and identities are always the result of processes of identification. It is through insertion in a manifold of practices and language games that specific forms of individualities are constructed. The political has a primary structuring role because social relations are ultimately contingent and any prevailing articulation results from an antagonistic confrontation whose outcome is not decided in advance. What is needed is therefore a strategy whose objective is, through a set of counter-hegemonic interventions, to disarticulate the existing hegemony and to establish a more progressive one thanks to a process of rearticulation of new and old elements into a different configuration of power.

The hegemonic strategy of 'war of position' that I am advocating is clearly informed by a conception of space which, like the one advocated by Massey, acknowledges its dimension of multiplicity (see Massey, 2005). By asserting that space and multiplicity are co-constitutive and that our constitutive interrelatedness implies spatiality, Massey's conception allows us to scrutinise the nature of spatiality and to see it as a field of political engagement. Her notion of 'power-geometries' brings to the fore the spatial character of the hegemonic articulations which constitute the nodal points around which a given hegemony is established. The globalised space appears as always striated, with a diversity of sites where relations of power are articulated in specific local, regional and national configurations. This reveals the crucial spatial dimension of the strategy of 'war of position' which has to take place in many different social spaces. Indeed, the multiplicity of nodal points which configure different geometries of power call for a variety of strategies and the struggle cannot simply be envisaged at the global level or in terms of desertion.

Conclusion

It is important to realise that, besides relying on different conceptions of spatiality, the disagreements between the two approaches that I have presented also stem from the very different ontologies that provide their theoretical framework. The strategy of exodus, based on an ontology of immanence, supposes the possibility of a redemptive leap into a society, beyond politics and sovereignty, where the Multitude would be able to immediately rule itself and 'act in concert', without the need of law or the state and where antagonism would have disappeared. The hegemonic strategy, in contrast, recognises that antagonism is irreducible and that, as a consequence, social objectivity can never be fully constituted. Therefore, a fully inclusive consensus or an 'absolute democracy' is never available. In all its versions the problem with this immanentist view is its incapacity to give account of the role of radical negativity, that is, antagonism. No doubt, negation is present in those theorists, and they even use the term 'antagonism', but this negation is not envisaged as radical negativity. It is either conceived in the mode of dialectical contradiction or simply as a real opposition. As we have shown in *Hegemony and Socialist Strategy*, to be able to envisage negation in the mode of antagonism requires a different ontological approach, where the primary ontological terrain is one of division, of failed *unicity*. Antagonism is not graspable in a problematic that sees society as a homogeneous space because this is incompatible with the recognition of radical negativity. In order to make room for radical negativity, we need to abandon the immanentist idea of a homogeneous saturated social space and acknowledge the role of heterogeneity. This requires relinquishing the idea of a society beyond division and power, without any need for law or the state and where in fact politics would have disappeared.

In fact, the strategy of exodus can be seen as the reformulation in a different vocabulary of the idea of communism as it was found in Marx. Indeed, there are many points in common between the views of the post-operaists and the traditional Marxist conception. To be sure, for them it is not any longer the proletariat but the Multitude that is the privileged political subject; however, in both cases, the state is seen a monolithic apparatus of domination that cannot be transformed. It has to 'wither away' in order to leave room for a reconciled society beyond law, power and sovereignty.

If our approach has been called post-Marxist, it is precisely because we have challenged the type of ontology subjacent to such a conception. By bringing to the fore the dimension of negativity which impedes the full totalisation of society, we have put into question the very possibility of such a reconciled society. To acknowledge the ineradicability of antagonism implies recognising that every form of order is necessarily a spatialised hegemonic one, that it constitutes a 'geometry of power', to use Massey's vocabulary. Heterogeneity can never be eliminated and antagonistic heterogeneity points to the limits of constitution of social objectivity. As far as politics is concerned, this points to the need to envisage it in terms of a hegemonic struggle between conflicting hegemonic projects attempting to incarnate the universal and to define the symbolic parameters of social life. Hegemony, as I have argued, is obtained through the construction of nodal points which discursively fix the meaning of institutions and social practices and articulate the 'common sense' through which a given conception of reality is established. Such a result will always be contingent and precarious and susceptible of being challenged by counter-hegemonic interventions. Politics always takes place in a field crisscrossed by antagonisms, and to envisage it exclusively as 'acting in concert' leads to erasing the ontological dimension of antagonism (which I have proposed calling 'the political') which provides its quasi-transcendental condition of possibility. A properly political intervention is always one that engages with a certain aspect of the existing hegemony with the objective of disarticulating/rearticulating its constitutive elements. It can never be merely oppositional or conceived as desertion because it aims at rearticulating the situation in a new configuration.

Another important aspect of a hegemonic politics lies in establishing a 'chain of equivalences' between various demands, so as to transform them into claims that will challenge the existing structure of power relations. It is clear that the ensemble of democratic demands that exist in our societies do not necessarily converge and they can even be in conflict with each other. This is why they need to be articulated politically. This is missed by the various advocates of the Multitude, who seem to believe that it possesses a natural unity which does not need political articulation because it already has something in common: the general intellect.

What is at stake is the creation of a 'we', a 'People', and this requires the determination of a 'they'. Virno's rejection (shared by Hardt and Negri) of the notion of the People as being homogeneous and expressed in a unitary general will which does not leave room for multiplicity is totally misplaced when directed to the construction of the People through a chain of equivalence. As we have repeatedly emphasised, in this case we are dealing with a form of unity that respects diversity and does not eliminate differences, otherwise it would not be a relation of equivalence but a simple identity. It is only as far as democratic differences are opposed to forces or discourses that negate all of them that these differences can be substituted for each other. This is the reason why the construction of a 'People' requires defining an adversary. Such an adversary cannot be defined in broad general terms like 'Empire' or subsumed under an homogeneous label such as 'capitalism', but in terms of nodal points of power that need to be targeted and transformed in order to create the conditions for a new hegemony. It is a 'war of position' (Gramsci, 1971) that needs to be launched in a multiplicity of sites. This can only be done by establishing links between social movements, political parties and trade unions, as Doreen Massey's own political interventions have strived to do. To create, through the construction of a chain of equivalence, a 'collective will' aiming at the transformation of a wide range of institutions so as to establish new geometries of power is, in my view, the kind of critique suited to a radical politics.

Note

1   The terms 'operaismo' and 'post-operaismo' are terms used to refer to the autonomous workers' struggles of the 1970s, particularly those in Italy with which Antonio Negri and Paolo Virno were strongly associated. They have been critiqued for their exclusive accent on workers' struggles.

Chapter Two
Theorising Context

Lawrence Grossberg

Context is a key category in contemporary social and cultural analysis, and crucial to the unique brilliance of the work of Doreen Massey. Yet it is almost never defined, and even more rarely theorised.[1] In fact, one is confronted by a chaos of contexts, both empirically and conceptually, clearly related but usually in unspecified ways to notions of place, the local and locality. Even the sophisticated theorisation of places in the work of, for example, Escobar (2001) and Raffles (1999) leaves the relations unaddressed:

> Locality is both embodied and narrated and is, as a consequence, often highly mobile: places travel with the people through whom they are constituted. Locality, then, should not be confused with location. It is, rather, a set of relations, an ongoing politics, a density, in which places are discursively and imaginatively materialised and enacted through the practices of variously positioned people and political economies. (Raffles, 1999: 324)

Most discussions of context do not acknowledge two conflicting assumptions: first, context is spatial, defining a bounded interiority, a stable island of ordered presence in the midst of an otherwise empty or chaotic space; second, context is relational, constituted always by sets and trajectories of social relations and relationalities that establish its exteriority to itself. As Massey (2004: 11) asks: 'If the identities of places are indeed the product of relations

Spatial Politics: Essays for Doreen Massey, First Edition.
Edited by David Featherstone and Joe Painter.
© 2013 John Wiley & Sons, Ltd. Published 2013 by John Wiley & Sons, Ltd.

which spread way beyond them (if we think space/place in terms of flows and [dis]connectivities rather than in terms only of territories), then what should be the political relationship to those wider geographies of construction?'

A lot of contemporary social and cultural analysis externalises these two notions into its description of the contemporary context as a conflict between the local and the global, where the latter is both a political economic project (neoliberalism) and a spatial configuration. Despite the gains that such descriptive structure make available, there are serious, even fundamental, problems with it. It has too often revived a kind of economic reductionism, whereby everything can be explained by an economic bottom line. Additionally, everything is neatly slotted into a simple binary logic of the local and the global, which is taken as both spatial and temporal. Even more sophisticated versions rarely avoid these two problems. These logics, which continually re-emerge despite arguments by the likes of Stuart Hall (1991) and Doreen Massey (1997a), result in an 'ethnographical imaginary' that celebrates the hybridity of the 'local' (glocalisation) as the solution to the very binarism that created the problem in the first place. In fact, hybridity is all there is. Not only is the global as hybrid as the local, but the past was as hybrid as the present. If the past, the other, was never as simple or homogeneous or local or unified as we imagine, the present is probably not as fractured or heterogeneous or global as we assume.

Just as importantly, even a cursory glance of the literature on globalisation makes clear that there is a palpable undecidability about exactly where the question is, what sort of data are relevant, and how to interpret them. Globalisation can refer to planetary inclusiveness, the compression of the world, 'the stretching out of the geography of social relations', the transnational reach of processes and power, a new formation of power, the increasing rapidity of communication and transportation across space, and so on (Grossberg, 1997, 1999). The list of changes that globalisation involves includes changing institutions, regulatory structures, forms of agency, scales (both economic and culture), structures of identity and belonging, time–space compression and so forth. While I do believe that there is some truth in all of these, there is little guidance provided for figuring out what it is.

I would suggest that many discourses on globalisation are limited by their failure to theorise the concept of context, often conflating different notions and thus equating elements that operate in very different scales and dimensions. Many of the leading models of critical analysis and even cultural studies can be understood as practices of radical contextualisation: Marx's practice of historical specificity; Foucault's analysis of *dispositifs* and discursive apparatuses; pragmatism's sense of situated knowledges and actions; and Deleuze and Guattari's theory of the production of milieus, territories and strata. But such theories have rarely been described in these terms, and their notion of context rarely elaborated.

I propose theorising the concept of context as a singularity that is also a multiplicity, an active, organised and organising assemblage of relationalities (Hacking, 2004) that condition and modify the distribution, function and effects – the very being and identity – of the events that are themselves actively implicated in the production of the context itself. Contexts are produced even as they 'articulate' the 'facts' or individualities and relations that make them up. Contexts are always in relations to other contexts, producing complex sets of multidimensional relations and connections. They are the result of and embody multiple technologies – residual, dominant and emergent – that are actively engaged in the (self-) production of the context. These technologies define the mechanisms and modalities of articulation or becoming – the condensations, of multiple apparatuses, multiple processes, multiple projects and multiple formations – that impose a particular organisation, individuality and conduct on the 'populations' of the context (Deleuze, 1988).

But if we are to theorise the complexity, the multidimensionality, of contexts, then we have to begin with questions about space itself. Contemporary understandings begin with Lefebvre's (1991) arguments that space is both made (emergent) and given (real). Massey (1992) extends this to argue that space has a density (substance) of its own, that space is constituted through flows and interactions, that it is the unfolding of interaction. As such, space is the very possibility of the existence of a simultaneous (positive) heterogeneity or multiplicity.

We can take this one step further: space is active and dynamic, the agency of the happening of multiplicity as possibility. Merleau-Ponty (cited in Casey, 1997: 296) sees space not as 'the setting (real or logical) in which things are arranged, but the means whereby the positing of things becomes possible'. Foucault understands space 'as specific forms of operations and interactions' (Kristin Ross, cited in Massey, 1993a: 67). And Deleuze and Guattari (1987) see space, not only as relational and active – the medium of becoming, as it were – but even more, as the capacity to actualise multiplicities, to produce relations or differences. I can elucidate this idea by appealing to the difference between four-dimensional Minkowskian space (in the specific theory of relativity) and the Riemannian space of the general theory of relativity. In the former, events are located and localised; they are points, with no power of their own even while they are independent of where they are. In the latter, events spatialise; they are 'world-lines', the vectors of the becoming of place. Space is not independent of what happens, and what happens is not independent of the space where it happens: a line can 'bend space around itself'.[2] Space is the movement of becoming itself and any 'place' has a power of its own. The very becoming of any event is the becoming of space itself. Massey once used the image of the plant blossoming to capture this sense of the co-becoming of event (relations) and space (personal conversation, 30 March 1998).[3]

I want to suggest that there are at least three ways of constituting contexts, three modalities of contextuality, three logics of contextualisation: milieu (or location), territory (or place) and region (or ontological epoch). They describe interconnected dimensions of every context, although the nature of that interconnection (e.g., a hierarchical/scalar relation) is itself contingent. Hence one cannot read the specific logic of one dimension onto or off another. They also describe ways of selectively mapping contexts. The best map is not always the articulation of all three together. How one maps a particular event/context will depend upon the problematic that is at stake. As Massey (2005: 175) puts it, 'the real political necessities are an insistence on the recognition of [the place's] specificity and an address to the particularity of the questions they pose' but without a 'vision of an always already constituted holism'. In fact, Foucault (n.d.) differentiates the radical contextuality of his 'nominalist' practice from both realism and social constructionism as problematisation:

> When I say I am studying the 'problematisation' of madness, crime, or sexuality, it is not a way of denying the reality of such phenomena. On the contrary, I have tried to show that it was precisely some real existent in the world that was the target of so much real discourse and regulation at a given moment. The question I raise is this one: How and why were very different things in the world *gathered* together, characterised, analysed, and treated as, for example, 'mental illness'? What are the elements which are relevant for a given 'problematisation'? ... The problematisation is an 'answer' to a concrete situation which is real.

Hence when considering some aspect of existence – for example, religion or derivatives or popular culture – one must carefully disentangle and distribute the questions into the appropriate dimensions. Many analyses of globalisation presuppose not only where the question is located but also direct relations of determination among the dimensions. Many analyses of the contemporary world conflate the different configurations of contextuality. For example, they equate the material processes and structures of power (milieus), and the embodied ways they are lived (territories). Or they substitute an ontological analysis (region), these days often Deleuzean, for the empirical description of the milieus and territories: they assume that a rhizomatic ontology guarantees that rhizomatic nature of the territory or that 'flat' (immanent, horizontal) ontology denies the empirical reality of verticality (e.g., scale).[4] A self-reflective theory of contexts – and an adequate contextual analysis – will have to theorise not only these different dimensions or modalities, but also the articulations among them. This is, I believe, an effort that is parallel to Massey's (2007: 191) own effort – 'to reimagine the relationship between flow and territory: to propose, in other words, another spatiality'.

Let me then try to describe each of these three modes of contextualisation. I will begin by briefly considering Deleuze and Guattari's (1987) distinction between the space of the milieu and that of the territory. In the first instance, space is a multiplicity of (overlapping) milieus as heterogeneous blocks of space-time. Milieus are the sum of the material relations within a particular space. Every milieu exists in complex spatial relations with other milieus: for example: 'the living thing has an exterior milieu of materials, an interior milieu of composing elements and composed substances, an intermediary milieu of membranes and limits, and an annexed milieu of energy sources' (Deleuze and Guattari, 1987: 313). Such contexts are not entirely random or chaotic; they are 'constituted by the periodic repetition of the component' or element. The boundaries of a milieu are defined by its material regularities.

Territories exist when there is a resonance or rhythm that articulates, coordinates or communicates across milieus, so that aspects or portions of the different milieu-contexts come together at a different level from the milieus themselves. The territory holds together some of the heterogeneous elements of already heterogeneous milieus, creating a kind of consistency: 'There is a territory precisely when milieu components cease to be directional, becoming dimensional instead, when they cease to be functional to be expressive ... What defines the territory is the emergence of matters of expression (qualities)' (Deleuze and Guattari, 1987: 315). Territories have a different mode of existence than milieus for they signal the emergence of matters of (non-subjective) expression. Such territories are created by everything from the song of the bird to the rites that found a city.

The identity of the territory is not defined by its inside alone, nor does it simply negate its outside. Expression constructs porous and mobile boundaries, an inside (of 'impulses' and activities) and an outside (of 'circumstances') and, in the process, it reorganises functions and regroups forces within the milieus. A territory is a consolidation across milieu-contexts, a holding together of heterogeneity by the expression of a rhythm among the elements. It is not a bit of space-time but an articulation across space-times to produce something else. Its interior is a dynamic site for carrying out actions and producing a place and a sense of belonging (an abode); it opens onto other territories and milieus, making it a space of passages and relays. It is an interiority that is inseparable from its outside, because the outside is only that onto which the boundary opens. A territory always has 'the interior zone of a residence or a shelter, the exterior zone of its domain' (Deleuze and Guattari, 1987: 314): A territory cannot be separated from the directional vectors of the milieus and the dimensional – expressive – resonances that move across milieus; it is neither origin nor destination. It is the organising of a limited space, a dynamic site for carrying out actions, a way of constantly holding back and opening up to the chaos, which is never only chaotic since it is also the space of milieus. 'How very important it is, when chaos threatens, to draw an inflatable, portable territory.'

It is not hard to see how these two spatial contextualities exist on the level (strata) of social existence. The milieu, or what I will call, in analytic terms, the location, describes a 'sociological' context, a material and discursive assemblage of political, economic, social and cultural practices, structures and events. But it is not simply what fills a bit of space-time; it is the very existence of that bit of space-time as the condition of possibility of what fills it, even as what fills it produces the space-time of the milieu. And it is constituted by the repetitions, the regularities, of the elements in the location. Indeed, Massey (2005: 175) introduces a similar notion of a location as 'an ever-shifting constellation of trajectories [which] poses the question of our thrown togetherness', that is, of our common socio-material existence in a common 'place'. Therefore we can think about locations together, in terms of the boundaries between them and the flows that cross them, although the identity of each location is *relatively* available on its own terms. But such relations introduce (vertical) scale – as extension – into a geography of contexts.

The territory – or what I will call, in analytic terms, and following Massey (2007), the place – is the context of lived reality. Importantly, the ways it connects to the material specificity of the location are always contingent, overdetermined and unpredictable.[5] The context as place describes an affective reality, or better, a complex set of affective articulations and registers – including configurations of investment and belonging, attention and mattering, pleasure and desire, and emotions. It involves the complicated relations between subjectivation and subjectification, emplacement and orientation, belonging and alienation, identity and identification. At its simplest, the territory describes the context as a lived topography. The territory is an expressive organisation of socio-spatio-temporal investments, transforming extensive space-time (the location), through intensive relations, into a liveable space-time. Places are different ways of living in already socially determined locations, different possibilities of the forms and configurations of belonging and identification, structuration and change, security and constraint, subjectification and agency. A place defines an orchestration of the affective tonalities that give resonance and timbre to our lives. It is clearly connected to Raymond Williams's (1961) notion of a structure of feeling. As Meaghan Morris (1992a: 467) suggests, a place is 'an organisation of the various time/spaces in which the labour, as well as the pleasure of everyday living is carried out'. It is an expressive and affective contextuality – marked by densities, distances and speeds – of access and agency, security and danger, mobility and networks, an assemblage of practices, discourses, experiences and affects.

Places have a different mode of demarcation; their boundaries are always unstable, fragile and porous, always somewhat indeterminable. In fact, one cannot think of the contextuality of places with a logic of boundaries; instead one needs to begin with a logic of connectivity that locates every place within 'webs of relations and practices'. Places only exist as always and already connected to other locations and places, as 'constellations of connections with strands reaching out beyond' (Massey, 2005: 187). They are contexts constituted by transits and translations. As result, they introduce (horizontal) scale – as intension – into a geography of contexts. Thus Massey (2007: 207) talks about places as 'meeting-places of multiple trajectories whose material co-presence has to be negotiated'. This 'reimagination of place' proposes

> a (potential) politics of place that looks from the inside out. It recognises not just ... The 'outside' that can be found within, but also – in a certain sense – the 'inside' that lies beyond. It poses the question of whether, in certain realms, we could 'live in the imagination' of 'our own place' as constituted through a distributed system – a kind of multi-locational place. (Massey, 2007: 193).

Finally, I want to consider a third mode of contextuality – the region as an ontological construction. A region is not a material location in space-time nor a lived place but the forms of existence – ways of being in space-time – that are possible and that constitute the contingent conditions of possibility of milieus and territories, locations and places. Ironically, as more people have turned to the highly abstract theoretical works of people such as Heidegger, Nancy, Agamben, Deleuze and Guattari and so on, the task of theorising context seems to have moved further and further away. I am not suggesting that the work of such philosophers is not crucial to this task but simply that it is not sufficient for the task. Moreover, too often, the sorts of hermeneutic or empirical transcendentalism offered in these works is substituted for the necessarily complex effort to offer better understandings of what is going on in particular contexts.[6] The ontological and the empirical are necessarily articulated, but they are also necessarily not the same; that is to say, they are always *articulated*! Ontologically, reality may be rhizomatic or flat; social existence may be conditioned by the inoperative community or the multitude; but these are hardly descriptions of the concrete relations of materialities and lived realities that constitute the complex contexts of social struggles and transformations. In fact, it is precisely the distance from the ontological that we have to measure, for it is here that power operates to produce the actuality of specific configurations of the ontological potentialities.

Yet ontologies of contexts are crucial in the attempt to theorise the context in ways that enable us to understand not only what is going on but also the ways contingencies have been realised and possibilities opened. Let me then briefly suggest some sense of what such ontologies have offered current critical work. The starting point of much of this ontological work is the hermeneutic ontology of Heidegger's *Being and Time* (1962), which enacts an analytic that moves from the ontic (empirical) to the ontological 'mode of being-in-the-world' of any being, including that sort of being that Heidegger calls Dasein (which includes the human being). Dasein is crucially constituted by/as a set of spatial and temporal relations and involvements. But it is only later in his writings, after attempting to dehumanise and desubjectivise ontology, that Heidegger offered an ontology of contexts or regions. The region is not only that which makes possible the being of any mode of being in the world, it is that which is given to us. It is a matrix of spatial-temporal possibilities, a structuring of involvements in which particular configurations of both locations and places can be specified, particularised and made intimate. In Heidegger's terms, the region specifies the possible ways in which 'Man' can 'dwell' in and with the world.

This ontology of dwelling finds its clearest expression in Heidegger's (1982) concept of 'epochs of Being' as a gift that is given to Man.[7] We are always only given a particular manifestation of reality that ontologically defines the possibilities of our dwelling, of the ways the world gives itself to us and of how we can organise and relate to the world. For example, the current (ontological) context, defined so completely by technology, is given in terms of the *Gestell* (enframing) and 'the world-picture'. The *Gestell* is an ontological region in which we as humans find our own Being, as well as the Being of beings, as a particular mode of experiencing and relating to the world as a spatio-temporal reality, in this case as resources to be used and used up. For Heidegger, humans do not create the epoch and they cannot choose to end it. We can only inquire: 'Might space have been determined elsewhere?' (Casey, 1997: 254). The point I want to make is this: while one may not adequately grasp the contemporary contexts of human life – the possibilities and limits of locations and places that have been determined elsewhere, as it were – without an understanding of the epoch or region we have been given, that ontological context is far from a useful or adequate description of the contextual realities of human life.

Certainly the most influential ontology of contexts is that offered by Deleuze and Guattari's theory of the becoming-actual of the real, which, while built upon Heidegger's contributions, attempts to avoid the anthro-centrism and semio-centrism so characteristic of contemporary theory, and to provide the ontological foundations of change itself. They start with the assumption that reality has two modalities of existence, both of which exist on a single plane; they refer to these as the planes of consistency and organisation. The former, the virtual, is a realm of unrealised but realisable capacities to affect and be affected (which they distinguish from the possible, which is not real). On the plane of consistency reality is the substantial multiplicity – real but not necessarily actualised rhizomes – comprised of lines of intensity or becoming, which are in fact both spatial and temporal.

But the plane of consistency is always and already organised *on the same plane*; a particular configuration of reality is actualised – produced – by the operation of multiple and specific machines or technologies. These machines create and organise populations and impose regimes of conduct, agency and effectivity on them. Such an actual reality, while ontologically flat, is also articulated into and across many different plateaus (e.g., inorganic, organic, human, etc.), each of which is organised by three kinds of machines (or machinic transformations): stratifying, coding (inscribing) and territorialising. These in turn embody three forms of relationality or articulation – connective, disjunctive and conjunctive respectively.[8] Each plateau or level of an actual reality is produced by and through processes that simultaneously stratify it into two assemblages – which Deleuze and Guattari call expression and content – or two populations and their forms of conduct. The first – expression – is characterised by 'functional or transformational' events, that is, by populations engaged with forms of activity and agency; the second – content – consists of populations which, on the plateau, give themselves to be perceived and/or acted upon. Such populations enact their own self-givenness or self-evidence. For example, on the plateau of the human, if the former describes the forms of perception and discourse, the latter describes that which is perceivable and say-able. Additionally, each of these populations is then further acted upon by two machinic processes: they are coded or inscribed with grids of differentiation and territorialised or distributed.

This ontology, all too schematically presented here, offers critics two analytic strategies (or tools): the first is a kind of deconstructive strategy that dismantles the plane of organisation, the specific configuration or actualisation of the virtual, to get back to the virtual as it were: we can always discover the rhizomatic, the flat ontology, the plane of consistency (immanence) as both the condition of possibility of and enacted as an excess inside the actual. Thereby, we can demonstrate that the real is always more than the actual; potentialities exist within – on the same plane as – the actual. This strategy is crucial if we are to hold together the idea that any empirical reality is a construction (reality produces itself machinically) and, simultaneously, that it is a contingent, stochastic outcome. Any struggle to change the world – even in ways that we know we cannot control – must begin with the understanding that the world does not have to be the way it is or better, that the world is already other than it is.

Although this is perhaps the most common appropriation of Deleuze and Guattari, I think the second strategy is both more valuable (at least for the kind of cultural studies I want to practise) and, in fact, is the precondition of the first tool if it is to have any concrete efficacy. This second strategy involves the analysis of the particular technologies or machinic processes by which a concrete actual reality is produced and sustained, often in ways that make it appear to be inevitable and exclusive. This involves the work of contextual analysis – whether or not one is a Deleuzean – and the moment in which an ontology of contexts demands to be complemented, on the one hand, by theories of both locations and places, and, on the other, by the actual empirical work of describing what is going on (as a production of power).

Let me offer one final example of an ontology of contexts that may help us to see why I have gone through the rather arcane exercise presented here. Consider Jean-Luc Nancy's (2007) contrast between globalisation and 'mondialisation' or world-creating. For Nancy, following Heidegger, the former marks the triumph of an economic and technological, or more generally, representational logic, producing a uniformity – 'a totality grasped as a whole' (Nancy, 2007: 20) – that can only result in injustice and an unlivable world. Under globality, 'the world has lost its capacity to "form a world"' (Nancy, 2007: 34). Thus, 'what forms a world today is exactly the conjunction of an unlimited process of eco-technological enframing and a vanishing of the possibilities of forms of life and/or of common ground' (Nancy, 2007: 95).

The latter – world-creating – refers to an 'authentic' world, the opposite of globality, a process rather than a totality, a space of commonality, signification and possibility. This mondiality is, in Nancy's terms, an absolute immanence, without the possibility of transcendence implied by the representational logic of globalisation. There can be no exteriority to such a world; it is the place of 'possible habitation … the place of possible taking-place … The world is the place and the dimension of a possibility to inhabit, to coexist' (Raffoul and Pettigrew, 2007: 9–10). The world is then not only a place, a habitus and a practice; it is also an ethos, an ethics of being in the world.[9]

While Nancy, in a 'prefatory note' allows that the conjunction – 'The creation of the world or globalisation' – 'must be understood simultaneously and alternatively in its disjunctive, substitutive, or conjunctive senses' (Nancy, 2007: 29), almost all of the affect of the writing pushes a disjunctive reading where the two modalities of worlding are opposed to one another (as an ethical matter). And this is how Nancy's theory is used most commonly, but it seems to me that there is a fundamental error here, contrasting, in Deleuze and Guattari's terms, mondiality as an ontological or virtual reality, and globality as a particular – and we might agree, destructive – actualisation. In fact, I would go further and say that while globality describes the world as a location/place, the very habitability of the mondial means that it describes the potentiality of places rather than an actual place, or actual ways of being-in-the-world. Nancy's ethics imagines the possibility of a place that is, somehow, self-reflexive about it own contingency and, therefore, is open to and even committed to its own creative self-transformation. If we do not distinguish these two concepts, we will find ourselves not only unable to distinguish and judge the various forms of globalisation, but also unable to see the possibility of change that is implicit in the recognition that globality is an articulation of – and therefore remains a form of – world-creating. And consequently, aren't all forms of world-creating also forms of globalisation in so far as they are actualised?

My point is not to challenge Nancy's ontology of the global but rather to argue, with Massey (1997a), that 'the very generality' of the ontological abstraction

> obscures the fact that ... There can never be 'globalisation in general' – if the world is becoming more interconnected then it is doing so, and must do so, in the context of particular power relations, and governed by particular political trajectories. ... The problem is that the generalised discourse of globalisation hides the fact of this politico-economic specificity. And it also, in consequence, hides the fact that there might be other terms on which the world's economies (and thus peoples) might be imagined. (Massey, 1997: 8)

A critical analysis of contemporary globalisation cannot be accomplished solely by understanding its existence as or productivity of some set of ontological contexts (e.g., Empire, or fractal postmodernisation). Nor can it be approached solely in terms of a reorganisation of spaces and places (the lived geographies), through an ethnography of how the global is lived in any place, or even in the global as a place. Nor can it be understood solely as a reorganisation and redistribution of the material and social locations that constitute the relational geographies of the world. It is in fact only when we have undertaken (i) the theoretical work of understanding each of these categories – locations, places and regions – and (ii) the empirical work of describing the forms and technologies of their actualisations, that we can then understand how they are articulated together into the 'teeth-gritting harmonies', the multiple, overlapping and complex totalities of social life and power. Let me, then, give the last word to Massey: 'In every age there is a making and re-making of the spaces and places through which we live our lives: what need to be addressed are the power relations through which that restructuring takes place' (Massey, 1997: 11).

Notes

1 This essay is drawn from Grossberg (2010).
2 I am appropriating the language of Bruno Latour and actor–network theory. Physicist John Wheeler observed, 'matter tells space how to curve, and space tells matter how to move'. http://map.gsfc.nasa.gov/universe/bb_theory.html
3 Despite using mathematics, I am opposed to the increasing willingness of social and cultural theorists to turn to contemporary scientific theories (complexity, chaos, networks) and to treat them not merely as tropes of imagination, but as sources of legitimation by virtue of their scientificity. The concepts have a long history in the humanities and are not suddenly better because scientists have come around to our ways of thinking. It would be nice if scientists would acknowledge our work, and admit they are discovering what the humanities have known for some time – or if they would share some of their grant support.

4 See, for example, the argument of Marston, Jones III and Woodward (2005). Their argument seems to me to involve a misreading of the Deleuzean flat ontology, which does not deny verticality, but argues that the molar and the molecular, the planes of consistency and organisation, the content and expression, all exist on the same ontological plane. Thus, it is also common in some readings to ignore the necessary coexistence of the plane of organisation and the plane of consistency.

5 This gap was made visible in cultural studies, for example, in the 'encoding-decoding' model, where it was argued that neither discursive nor sociological realities determine or guarantee the lived realities of what people make of and do with particular media texts.

6 I would add that they are neither universalist nor essentialist. They offer 'historical', or better, contextual ontologies.

7 This was, I believe, later taken up – albeit significantly revised – by Foucault.

8 I am aware that identifying the forms of machine with the forms of lines is at best an oversimplification but I nevertheless think it is a useful one.

9 Like Heidegger, Nancy does not think the origin of any world can be explained – it is neither necessary nor contingent, but simply a singularity, which for Nancy always includes an excess – that other beginnings, other worlds, are possible.

Chapter Three
Power-Geometry as Philosophy of Space

Arun Saldanha

Introduction

Why is geography indispensable for thinking capitalism and identities? Doreen Massey's work has been indomitable in positing the categories of space and place as fundamental to social inquiry. Taking my own experience of being Massey's doctoral student as illustration, this chapter understands Massey's oeuvre as 'philosophical geography', that is, as not just empirical social science but as a contribution to the production of concepts. Specifically, I will argue that Massey's concept of power-geometry is more than descriptive of modern global processes. Grounded as it is in important contemporary philosophical systems such as those of Marx, Bergson and Heidegger, and like other philosophical notions (justice, essence, practice), the concept of power-geometry aims to be as general as possible. It can be seen as coterminous with complex society, or even the human species itself. The chapter will close with some questions, derived from Althusser, about the ambivalent relation Massey has with the Marxist legacy.

Space as Open and Closed

Although almost every Briton knows the Open University (OU) from late-night BBC2 television, few have heard of its campus in Milton Keynes.

Spatial Politics: Essays for Doreen Massey, First Edition.
Edited by David Featherstone and Joe Painter.
© 2013 John Wiley & Sons, Ltd. Published 2013 by John Wiley & Sons, Ltd.

From the 1950s, plans were formed to build a city of 250,000 next to the M1 motorway to Birmingham, to alleviate the density, congestion and poor housing of London. Similar to other mid-twentieth-century projects such as Brasilía and Canberra, the plans for Milton Keynes followed social-democratic beliefs in progress and equality, which also underpinned the conception of the OU.

As can be expected, geometrical rationality reigns in Milton Keynes, as does an investment-friendly mixture of public and private (Allen *et al.*, 1998). The city's unique grid structure of quasi-highways and roundabouts allows cars to easily negotiate the 1 km squared estates, with clear zones for residence, shopping, business, education and leisure. Existing villages, fields and market towns have been absorbed by what may now be the fastest-growing metropolitan area in Europe, its population having almost approached the projected quarter million. The city encompasses what could be called the region's original science park with its concomitant masculinist 'high-tech fantasies', as Doreen Massey labelled them (Massey, Quintas and Wield, 1991): Bletchley Park, where Alan Turing himself helped break the Enigma code of the Nazis.

A spatial conception of how a city is constituted through planning, war, transport networks, immigration and retail does not come naturally. Being trained in communication and cultural studies in Brussels, it gradually dawned on me that the study of social meaning could not do without a rigorous theorisation of space and place. I discovered that Doreen Massey's geography both triggered and encompassed many central themes in the cultural studies paradigm associated with Stuart Hall: identity, globalisation, travel, hybridity, patriarchy, politics. Fortunate to receive a studentship at the Open University to continue my doctoral work on rave tourism in Goa, Doreen and Jenny Robinson became my supervisors.

It is no accident that Doreen worked, like Hall, for the Open University, which will bear their stamps for decades. She and the university both embody and advocate pedagogical values such as interdisciplinarity, life-time learning and international exchange. The OU has its roots in the egalitarianism of the BBC and the Labour Party's commitment to addressing the deep class divisions in the British education system. Over the 1960s the idea took shape to found a university without lecture halls, which would use electronic media and books instead to open up the ivory tower of knowledge. The rapid success of the project – 70,000 students graduated by 1980 and 200,000 by 1998 – shows that the OU definitely fills an important gap left by post-industrial capitalism. As Raymond Williams, intellectual precursor to Hall and Massey and strong supporter of adult education, noted in the late 1970s, the OU was never set up to actually change class inequalities (1979: 371). Under neoliberalism especially, companies gladly support continued education if their workforce becomes more productive. The OU has partnered with business education institutions worldwide, and its campus in Milton Keynes has in many ways become an external grant-funded science park. Still, it is no exaggeration to say that the OU drastically transformed the relationship between the geography of knowledge and the geography of class.

Massey (2005) understands space as an emergent physical, affective and political reality, traversed and constituted by flows and hence continuously changing. The boundaries of a place are made and remade through corporeal practices (walking), representations (the local newspapers) and the work of objects (fences). It is not so much local history that defines what a place 'is', but its connectedness to the outside world. There are huge differences between the amount and kinds of connectedness that places harbour: a multicultural finance hub like London at one extreme and a farm on the Hebrides at another (to remain within the UK). But the dependence of places on long-distance networks often comes as a surprise if we think of them as simply local. Within places, Massey furthermore argues, different people have various degrees of access to the flows that go into making their place. Men in particular have a lot more freedom of movement and public presence than women.

Relating her commutes with Stuart Hall, Massey writes:

> when you make that journey from London to Milton Keynes, you are *not* just travelling across space. ... You are part of the constant process of the making and breaking of links that is an element in the constitution (1) of you yourself, (2) of London, which will not have the pleasure of your company for the day, (3) of Milton Keynes, which will (and whose existence, as, say, an independent node of commuting is reinforced as a result) and thus (4) of space itself. ... And then, come the evening, weary, we set off again, making our way home to the big city. Yet that going home is not at all going back to the same place. London is not the same place we left this morning. (2000b: 225–227)

A world city, London is not so much a microcosmic resemblance of globalisation (to start with, the globe has oceans), as a central instigator of flows of commuters, migrants, businesspeople, financial services and so on (Massey, 2007). It actively maintains the global disparities that are variously folded back into itself (most tragically in the terrorist attacks of 7/7). In providing a reservoir of labour, Milton Keynes is a key London satellite. Travelling between the two places is to further interweave them.

Of course, the tale is of many more than two cities: there's Birmingham, Oxford, Cambridge, Brussels, Frankfurt, New York, Nairobi, and so on. Places are, for Massey, constituted relationally through the links afforded by the railway, motorway, telephone, as well as many kinds of imaginations. Places are by ontological necessity historical, continuously reshaped through what Massey (2005: 13) calls the 'throwntogetherness' of trajectories of things and bodies. That places have no 'essence' does not mean they have no internal history and agency. Massey's famous global sense of place (1994a: 6) is precisely an affirmation of such an inner place-reality as product of the place's multiple relationships with elsewhere. Milton Keynes itself materially changed every time Massey and Hall travelled there, just like it did with every architect, bird or pop star passing through it, however slightly. Travellers do not just simply arrive at and leave places, of course, and circulate within those places in different ways. All of space is produced differentially through the various social groups that inhabit and alter it from within.

The mobility of corporate and diplomatic elites or refugees supplies them with more intense mental mappings of globalisation. Continued residence means having a denser mental mapping of a place. If few of the Open University's academic staff live in Milton Keynes itself, the main reason for this can be inferred from my brief description of the city's modernism above. Most who have visited the city will agree that few others match its strange scatteredness. Although it contains picturesque pockets of early modern, even medieval rural England, Milton Keynes as a whole is almost impossible to explore on foot. Its sterile housing, roundabouts, underpasses, giant malls, lawns, parking lots, an indoor ski slope and a famously metonymic public art piece of concrete cows can be bemoaned aesthetically, but these are exactly what makes the city attractive to businesses and working-class families (Allen et al., 1998: 84–89). More important, however, is that this landscape indexes the systematic removal of genuine public space under neoliberal capitalism. As a city Milton Keynes does not have the thick layers of history that otherwise afford a socio-political loyalty to place.

It is easily explained, therefore, why many at the OU prefer working at home. Unlike in other cities, there can be no spillover into pubs after work. Meanwhile, from the viewpoint of Milton Keynes's inhabitants, there is little that would bring them to campus. Hence there is a quintessentially geographical paradox: even if its 'virtual' pedagogy makes the OU one of the most *open* of educational institutions in the world, its physical grounding in Milton Keynes makes it unintentionally *closed* to the proximate world. This is not because of its security policy or its buildings or a shyness of its employees, but simply its location.

To summarise, Massey's example of the London–Milton Keynes commute evokes key concepts in Continental philosophy: space and time, dwelling, memory, capitalism, sociability, social justice. Places are not homogeneously integrated units. Within Milton Keynes, the 'OU community' is mostly barred from the politics of urban space. To commence philosophical geography, the physical closedness of a virtually open university is perhaps the right sort of conundrum.

Doreen Massey's Philosophical Geography

Following Deleuze and Guattari (1994), philosophy is not empirical but deals with abstraction, while the sciences, whether human or physical, are committed to understanding bits of the real world. What makes Doreen Massey so central to our discipline is that she is the most systematic *philosophical geographer*. From Kant onwards, the domain of philosophical anthropology has elaborated on the implications for Western thought of ethnographic and anatomical research, but its main traditional question 'What is man?' is now decidedly old-fashioned (see Foucault, 2008). What I call philosophical geography takes off where philosophical anthropology reaches its Euro-masculinist limits, by trying to formulate what philosophy can and should learn from the geographical archive on places, landscapes, countries and urban networks. The guiding question of philosophical geography is 'What is space?' In drawing from a diversity of philosophical sources to construct a new concept of space, Massey leaves the discourse of geographical science and carries out properly philosophical work that few institutional philosophers have cared about.

To conceive our discipline's bizarrely general object, space, Massey has had to invent a crossroads of a number of philosophical systems. However, I will argue that Massey manages to distil these at times disparate philosophical flows into her most evocative concept, *power-geometry*. Although other geographers have been equally influential in suggesting new concepts, it is either the narrowness or the eclectic abundance of their philosophical sources that prevent them from attaining the neatness of this synthetic, more-than-empirical concept. After discussing where the concept of power-geometry comes from philosophically, my third and last section signals some points for possible further discussion.

The concept of power-geometry is fundamentally about patterns of unequal relationships, and it is in Marxist political economy that Massey's oeuvre commences. Understanding that there can be no surplus without labour, Massey's landmark study of industrial restructuring in the UK, *Spatial Divisions of Labour* (1995b), demonstrates that capitalism does not just produce but requires bouts of regional unemployment. In foregrounding labour and practice the study can be seen together with contemporaneous efforts of some philosophers following Spinoza, such as Louis Althusser, Etienne Balibar and Antonio Negri, to extract from Marxism an appreciation of the open temporality of social formations and the multiple relationships intersecting in those formations (Hardt and Negri, 1994; Balibar, 2007; Althusser, 2005). Like Henri Lefebvre (1991), however, Massey insists on the primacy of the *spatiality* of production. With the elegant addition of the adjective 'spatial' in her title she argues that qualitative and quantitative differences between the skills and bodies available in certain places *cause* companies to invest or disinvest in them. Spatial structure is no mere outcome of social relationships but consubstantial with them. Bourgeois economics' promise that capitalism brings equilibrium is made only by doggedly removing spatial heterogeneity as a variable.

The switch in Massey's late-1980s work from spatial divisions of labour to power-geometry is firstly one of thematic scope. Power-geometries are made by much more than investment in jobs: 'To say that a society is capitalist is not to say that it is not other things as well, such as racist, homophobic, sexist' (Massey, 1995b: 299). But the switch is also an ontological and epistemological one. Massey mostly abandoned political economy as a starting point for social analysis. Turning to post-structuralist philosophy, especially perhaps Derrida, she condemns the residual positivism and phallogocentrism in Marx, unpacking the traditional analogies Western thought had spun between binaries such as time/space, male/female and culture/nature (see Massey, 1994a, for critiques of Marxist geography).

Apart from Marxism, another important adversary of the concept of power-geometry is phenomenology and its elaborations in humanistic geography. Heidegger's famous concept of authenticity is explicitly spatialised into a celebration of dwelling and a suspicion of technocratic modernity (Malpas, 2007). On Massey's understanding of power-geometry (1993a), any dwelling is instead already an effect of migrations, already a relay in networks of circulation, already tense with oppression – at the very least, usually at least, of women. The project of feminist geography that Massey helped inaugurate in the 1980s seeks to bring the questioning of sexual identity in the political realm to bear on what is traditionally known as the relationship between 'man' and world.

Further fuel for a critique of humanism and universalism comes from Althusser's legacy in Ernesto Laclau and Chantal Mouffe. Despite Laclau's adherence to old prejudices against space (Massey, 1994a), the deconstructionist rejuvenation of progressive politics in Laclau and Mouffe (2001 [1985]) corresponds well to Massey's privileging of relationality above identity, radical openness above the maintenance of wholeness and origins. Massey (2005: 20–24) traces the prejudices against space to Bergson (2008), who held that time could not be measured and segmented like space without violating its proper dimension of becoming. 'Spatialisation' is in Bergson's early philosophy therefore equivalent to a fixing of becoming in representation, to death.

Massey reworks the Bergsonian concepts of multiplicity and becoming (identified by Deleuze, 1991, as two of Bergson's greatest contributions to philosophy) by obverting the denigration of space. It is simply impossible for a geographer to accept that spatialisation equals homogenisation.

Multiplicity has to be spatialised or it is a mere aggregate of elements, united only in consciousness. Hence:
- for there to be time there must be interaction;
- for there to be interaction there must be multiplicity;
- for there to be multiplicity there must be space. (Massey, 1999a: 33)

As in quantum physics Massey (1994a) understands space-time as an indivisible continuum (but see 2005). There can be no process without prior differentiality and heterogeneity, that is, without space. And vice versa, for things to emerge, 'fall into place' and interact, time has to elapse. This is more or less how Deleuze and Guattari (1987) understand multiplicity. Of all contemporary philosophy it is their 'geophilosophy' (1994) that arguably comes closest to Massey's philosophical-geographical project.

Massey's post-structuralist ontology of place therefore amends the Bergsonian tradition by showing that its conceptions about space are misplaced given its own logic. Bergson does not appreciate that all creativity, intensity and embodiment *take place*. And place is precisely de-essentialised, multiplicitous and productive because its connectedness constantly reignites temporality within. Because it is internally disrupted by time, space cannot be known fully in/as itself. There is no objective 'snapshot' possible of space, even of *a* space. Here Althusser's notions of dislocation and complex totality and related critique of Hegel's concept of time (Althusser and Balibar, 2009) have long inspired Massey (see Massey, 1997b; 2005). Space is the synchronicity of many entangled diachronic lines, which no one can freeze like some sort of God could, as if observing from without. More epistemologically, Massey joins feminist critiques of science and universalism by insisting that there is an inescapable situatedness of all observation, all theory, all views of time. If space is relational and relative to what occupies it, so is any concept of it.

Apart from Marx and Althusser, Heidegger and Bergson, post-structuralism and feminism, a last affinity of Massey's philosophical geography is the varied work that has taken place under the label 'complexity theory' (Prigogine and Stengers, 1984; see Massey, 2005; Harrison, Massey and Richards, 2006). Complexity theory's discipline-transgressing and now almost ubiquitous notions, such as self-organisation and emergence, have always been explicitly spatial. However abstract, the modelling of the dynamics of urbanisation or flocks of birds has to engage repeated interactions between components, hence multiplicity, hence space. Through the non-reductionist thinking of complexity theory Massey (2000b) inserts herself in a Spinozist tradition in which humans do not inhabit their own kingdom separate from the rest of nature. She also continues, though far from uncritically, geography's venerable tradition of conceptualising the interactions between earth, culture and mobility.

Power-topology

To the extent power-geometry is a philosophical concept I want to argue that it uncovers a fundamental aspect of reality. More than any other species, humans have been spinning increasingly far-flung and dense networks of exchange and environmental modification for at least a million years

(McNeill and McNeill, 2003). Power-geometry explains how patterns of population flows and interactions gradually intensified into today's obscene disparities of global capitalism. Survival is usually – ideologically – explained by local environmental factors, thereby downplaying the migrations, invasions, nomadism, long-distance trade, exploitation, pillaging and so on that constitute complex societies (Diamond, 1997, is the best known example). In densely populated regions like Eastern China, South Asia and Europe, power-geometry has for at least 10,000 years mediated all environmental stress and is thus the 'ultimate cause' of who in the region dies prematurely, and who lives a comfortable life.

In exposing interconnections and gradients of capacities between people, the term power-*geometry* can then be taken quite literally. In the middle of the nineteenth century, differential calculus and the ancient analysis of points, lines and planes in Euclid's *Geometry* were expanded by Bernhard Riemann (DeLanda, 2002). Mathematical spaces or manifolds could henceforth accommodate transformations along not just three but any number of dimensions. Riemann thus enabled a new branch of mathematics called topology, central to the later physics of space-time, cybernetics and today's theoretical biology. Like topology abstracts from physical space-time to think networks, knots and dimensionality *as such*, Massey's geometry thinks the intersecting differentials which economic, cultural and political forces impart between places and human bodies. Using words and maps instead of diagrams or algebra, Massey presents space in its manifold, continually changing nature.

In calling explicitly for a topological analysis of power's spatiality, Massey's colleague John Allen (2003) theorises the *modalities* of bureaucratic and corporate power. The less Euclidian our concept of power-geometry, it would seem, the more attuned it becomes to the complexities of real unequal multiplicities. We can map the population densities of southern England, the railway and electronic connections between London and Milton Keynes, their administrative boundaries, the investments and wages that pass between the two cities, and conduct surveys on customer satisfaction of commuters. This would describe the space the two cities share as quantitative, externalist, reified multiplicity, and can be convenient especially to capitalism. Power-geometry digs deeper, however, to see what quantitative multiplicity actively covers up, in order to *explain* space. Before it is measurable and mappable, space is topological, made up of structural disparities, gradients and affective connectivities. It is such abstraction that Althusser argues is central to Marxist science.

Here another concept that Deleuze (1991) retrieves from Bergson becomes valuable: the virtual. The tendencies, probabilities and patterns that shape matter are virtual in that they have their own ontological reality. Unlike potentialities in traditional ontology, their actual outcome is not determined in advance; unlike possibilities, they lack no reality (DeLanda,

2002). The imaginative power of post-Riemannian geometry for empirical science lies in positing and researching a virtual realm of transformations, twists and codependencies that actual entities are involved in or are capable of. This kind of space is what physicists call 'phase space', and biologists 'fitness landscapes', while Bergson is unfortunately reluctant to call it space at all.

I think Bergson's philosophy of time is wrongly opposed to 'space'. In fact, his concept of virtual multiplicity can complete the materialism that critical geography seeks. The geometry of human power relationships gains in robustness with a detailed notion of the real-but-not-(yet?)-material realm of the virtual. This robustness is both philosophical and political, since actual inequalities are submerged in a virtual sea of possibilities to change them. Perhaps Massey's critique focuses too much on Bergson's misunderstanding of space-time and too little on Bergson's ontology of the virtual constitution of the passing of time.

Nonetheless, given the enduring obsession with temporality in cultural and political theory, due in no small measure to a revival of Bergson, Massey's strong appeal to space remains indispensable. Furthermore, the philosophical richness of Massey's formulation of a *concept* of space never implied a retreat into armchair politicking, as at times happens in cultural and political theory, much less a quasi-mysticism à la Bergson. True to Marx, her theorising always commenced from and therefore facilitated her ongoing political engagements both locally and internationally. Hence unlike Bergsonian duration, the concept of power-geometry was, somehow, echoed in Hugo Chávez's new constitution for Venezuela in order to postulate a new geopolitics (Di Giminiani, 2007; Massey, 2008). If power-geometry is central to philosophical geography, politics has to be too.

For Marx

Massey's title *For Space* (2005) clearly evokes Althusser's *For Marx*, but in which way? The affirmations of Massey and Althusser differ on epistemological and historical grounds. Althusser (2005) is 'for Marx' by affirming the continuing latencies for politics and science that Marxist theory contains, an affirmation against both official and humanist-existentialist readings of Marx prevalent in the 1960s. Massey is *for* a theory of space as open, multiple, processual, and against evolutionist ideologies inherited from the nineteenth century, including developmentalist versions of Marxism, against space as parcelled, calculable, inevitable. While Althusser's polemical aim is to retrieve the originality of one foundational figure of the social sciences, Massey's is to elucidate a much-misunderstood philosophical category via an inclusive range of theoretical registers. While Althusser's 'for' is a declaration of creative fidelity, Massey's 'for' is an abbreviation for the breadth of her intellectual achievements regarding spatiality, after Marx seemed more or less depleted to her. But in concluding this chapter I would like to suggest that Althusser's and other recent rereadings of Marx prove that his relevance is not yet depleted. Affirming space still requires affirming Marx anew.

Massey's reluctances about Marxism are fully legitimate. Marxism has always placed too much emphasis on studying capital flow over the experience of labour; on rationalities over affects; on macho uprising against piecemeal efficacy. And yet *Spatial Divisions of Labour* made a highly original contribution not just to the geographical but to the philosophical conceptualisation of capitalism, with an exploration of industrial labour as *inherently* spatialised through its class, gender and racial differentiations. Combined with the introduction of a 'geological' notion of the accumulation of place-effects, there is nothing quite like *Spatial Divisions of Labour* in the Marxist literature. Even if it exceeded narrow Marxist theory, therefore, I would argue the book partakes in a wider ongoing project of recasting Marxism (through Althusser, Deleuze and Guattari, Gramscian cultural studies, Laclau and Mouffe, etc.) into the exigencies of the globalising present. The massive internal poverty and environmental disaster that accompany the capitalisms of China, India and Russia, and the cynicism of bankers in the City and Wall Street, indicate that spatial divisions of labour have become even more obscene than during the Cold War 1980s. This requires deepening our understandings of global finance flows and the masculinist regimes they are fuelled by.

A continuing philosophical relevance of Marx, and especially the labour theory of value, was championed by Negri in Italy from the late 1960s. Like the later Massey, Negri has strong affinities with Spinoza and Deleuze and Guattari. More explicitly than the contemporary British left, however, Negri embraces Marx's idea that communist subversion is rooted in the pragmatic reorganisation of work: 'The world is labour. When Marx posed labour as the substance of human history, then, he erred perhaps not by going too far, but rather by not going far enough' (Hardt and Negri, 1994: 11). For Negri, 'labour' (waged, housework, education) is in its abstractness the very same productive sphere that produces value for capital to grow *and*, always-already in many pockets and potentially globally, novel communist kinds of making and living that are antagonistic to capital and the state.

While Williams still declared his sympathy for Maoism (1979), the decided turn away from communist idealism within British academia can be understood within a more general redefining of the European left over the 1980s and 1990s under the pressure of the neoliberal right. The names of Marx and Lenin came to signify the hopelessly passé past. In abandoning the hope of thinking a viable alternative to capitalism's relentless colonisation of every sector of labour and leisure, Blairite liberalism effectively lost its every anchorage in a historical lineage of strife and political passion for social justice. It is against such loss that Massey's theoretical innovations and activism are directed, and have to be directed again.

More attuned to global complexities than Negri, geographers have rightly criticised his philosophical position for overestimating the coming regime of alleged 'immaterial labour' and the independent force of counter-globalist movements (e.g., Allen, 2003; Massey, 2005). But conversely, what the decline in reworkings of Marx in British cultural studies and geography risks is losing a quasi forever-relevant pool of concepts and political strategies. As neoliberalism and the war on terror combine into an onslaught on even the tamest achievements of multicultural and egalitarian democracy, creatively re-examining historical materialism and the failures of socialist revolutions is as timely as ever.

Conclusions

What may the relevance be of Marx to my brief geographical portrait of Milton Keynes above? The city was very successful in creating jobs and urban growth, courting Japanese, American and European companies (electronics, construction, banking, training) to relocate their headquarters there. Tens of thousands of working and consuming households could be concentrated close to a world city. But while the Open University maintains a democratic objective of distributing opportunity, the marked class and racial disparities between the Milton Keynes estates, and its fundamental belief in the nuclear family and private property, including of the means of transport, suggest accomplishments of capitalism rather than social democracy.

Now, few Milton Keynesians I met complained about alienation. As a student I appreciated the convenience of concentrated retail, the bicycle network, the quiet campus. That Milton Keynes became the location for attracting capital and labour out of London (including the Open University) is an all-too-easy target for critique. Still, the larger urban and ideological systems from which the city sprang – the UK's aggressive version of consumer capitalism and its co-optation of the Labour Party – do require sustained Marxian analysis. What the disjuncture between the OU campus and this 'New Town' lays bare are some fundamental contradictions in the 1960s' remaining Marxist ideals. The same ideals that produced a cold diffuseness made it subsequently difficult for a university upholding many of those ideals to develop a local presence.

It is not the case that Doreen Massey, Stuart Hall or anyone else at the OU has been unreflective about the ironic split between the university and the city or the co-optation that Williams pointed out. The reinvention of sturdy alternatives to capitalism seems as far away from a city like Milton Keynes as peace is to Israel/Palestine. But when every year brings the absurd unsustainability of inherited modes of doing things into yet more drastic relief in Tel Aviv as in Milton Keynes, London, Brussels, Nairobi, such reinvention also becomes more necessary. Marx has been the historical name for the analysis of the intrinsic pathway to disaster that is capitalism. Though in many places there certainly remains a stuffy macho feel to it, a re-engagement with the Marxian vocabulary and its search for social justice will most benefit from the contributions of those who first escaped it.

Acknowledgements

Thanks to the editors for suggestions on a previous draft and the Raymond Williams reference.

Chapter Four
Spatial Relations and Human Relations

Michael Rustin

Introduction

From the 1980s onwards, sociologists could with reason look enviously across at their colleagues in the field of geography, and ask themselves what had happened to bring about such a reversal of their respective intellectual and academic fortunes.

In the 1960s and 1970s, the discipline of sociology seemed to provide the most effective language with which to think about social change and conflict. Indeed it was probably because its own internal arguments and rifts were so focused at the time on the theoretical and political necessity to think about change and conflict that this became so. This was the period, after all, of the breakdown of the Keynesian, New Deal or Welfare State settlement in the United States and Britain, as the clamorous voices of class, race, gender and generation demanded a voice which each said were denied to them.

But then an intellectual revolution took place. For some reason, the sociologists, who had succeeded in becoming the leading force in charting and even giving political expression to the growing antagonisms of the era of industrial conflict, galloping inflation and 'ungovernability', seemed less able to function effectively as the theorists and catalysts of the era that followed. Neoliberalism after all reinstated the individual as the focus of moral value, and as the prime mover of change, through the roles of entrepreneur and consumer in which 'individuality' was held to be most

Spatial Politics: Essays for Doreen Massey, First Edition.
Edited by David Featherstone and Joe Painter.
© 2013 John Wiley & Sons, Ltd. Published 2013 by John Wiley & Sons, Ltd.

fully expressed. It was at this time, in the early 1980s, that geography as a discipline revolutionised itself, and seized its opportunity to become the most effective critical analyst of what was fast becoming 'the new world order'. 'Space' suddenly became a focal concept, as it had not previously been. Changes in the meaning and experience of space, and transformations in human relationships to it, became one of the most powerful metaphors for explaining what was going on.

Doreen Massey's work was, of course, central to this reinvigorated and transformed geography. She has been concerned to chart the impact of these larger spatio-temporal processes on the embodied human societies she knows and cares about. One cannot exactly say 'local' societies, since although she has retained strong attachments to certain specific places (her Manchester working-class origins, for example), she has also made and kept many other 'local' attachments in the course of her working life; for example, she has worked with people in several parts of Latin America, including Nicaragua, Mexico and Venezuela. She has also retained keen links to the Open University, with its distinctively democratic educational mission, and to London itself (and the particular neighbourhood within it where she lives[1]), where one of her most active political and indeed governmental commitments was her work with the Greater London Council and with the London Mayor's Office of Ken Livingstone. Her work has shown that it is possible to be both a severe critic of the consequences of a nexus of financial and governmental power – for example in bringing about inequalities between North and South in Britain, and at the larger but also interconnected level of neo-imperial relations – yet also enjoy and value the more benign side of the cosmopolitan diversity of a great metropolis. After interrogating Massey's account of space and time, this chapter explores possible affinities between her approach and the relational turn in variants of psychoanalysis.

The Categories of Space and Time

Massey in her work has given attention to many substantive forms of spatially located inequalities, such as those of North and South both within Britain and on a global plane, and in regard to inequalities of gender. But she has also extensively explored the abstract categories of space and time themselves, especially in her later work (Massey 1995a, 2005). While her work has been distinguished in its consideration of these issues (it has also increased in its complexity and sophistication over time), it has not been wholly untypical of work in the 'new geography'. This discipline seems to have found it necessary to interrogate its subject matter at this 'meta' level, in order to find ways of making sense of its substantive objects of interest. The surprising thing for me in this project is that the categories of space and time appear to constitute the philosophical presuppositions of a geographical form of study, but not to be conventional empirical or theoretical objects of study in their own right. Sociologists can study social groups, or forms of social action, or social networks, or social structures (as the sedimented forms of these), and still remain quite close to the observed particulars which constitute their empirical field. Something similar can be said of economists and their observable particulars of interest-maximising individuals or the competitive markets within which they interact; or of anthropologists and the normatively or culturally shaped patterns of behaviour which they study. The ontology of these fields moves between observed particulars and typified generalities and causal or other relations between both, without moving that far away from the tangible empirical objects which might be thought to constitute the usual objects of most kinds of scientific study.

But the concepts of space and time seem to function at a rather higher level of abstraction than this, that is to say philosophically as much as scientifically. Entities need to exist in both space and time in order to exist, but to say an entity does have these two dimensions seems to ascribe no attribute except existence itself to the entity in question. It is interesting therefore that a field of study should have decided to define itself, in recent decades, in part because of its distinctive preoccupation with these categories. What has been at stake in this choice, and in particular what has been at stake in it for Doreen Massey?

It seems to me that the insistence that, as a favourite maxim would have it, 'space matters', can only be understood in the context of its implied contrary. And if the implied contrary might seem literally to be that space was of no significance (hardly a conceivable idea at all), its real force was to refute an opposite and erroneous focus upon the dimension of time, to the exclusion of space. To whom was this opposite emphasis imputed? One might say to two different but related forms of historicism ('historicism' meaning the construction of deterministic narratives of history to which all and every societal particular was presumed to conform). One of these historicisms was the once-orthodox version of historical materialism, the idea that the succession of modes of production and of the hegemonic social classes emergent from them would and should determine the shape of the future. The other historicism was that of liberal capitalism and its associate forms of 'modernisation theory', embodied in recent years in the 'Washington Consensus' and for example in the premature declaration, after the fall of European communism, of 'the end of history' by Francis Fukuyama (1992). Doreen Massey is implacably hostile to the neoliberal project of development, especially in its effects upon the regions and countries of the developing world, Latin America in particular. And she also has little time for the theory of the inexorable rise of the working class, not least because of her and others' feminist understandings that the working class in this political reckoning was all too liable to trample over women during its forward march.

The absolute priority which had been accorded to the category of time in this reading was held to flatten out and homogenise all differences between social subjects, other than those relevant to their position in the imputed line of linear historical advance. The contrary – or in fact complementary – emphasis on the category of space signified a commitment to the recognition of differences, not only or mainly in regard to the imputed line of historical march (whether that be bourgeois or proletarian) but in regard to differences and particularities as such. Entities cannot occupy the identical space at the same time, and therefore in principle the insistence on the recognition of the innate multiplicity of spaces and places means an insistence on a principle of difference, and of differences as a source of value.

The argument, however, is not for an explanatory emphasis on space alone – Massey has been at great pains to insist on this point. Emphasis on the dimension of space, to the exclusion of that of time, is as crushing of complexity and variety as the reverse choice. If there is only space, and no dimension of time, there can be no change and no development. Just as if there is only time and no space, there is change only within a uniform and homogeneous social substance. In practice, predominantly spatialised representations of social reality take various forms. One of these is a fundamental conservative insistence that nothing of moment within 'our' spatially located community (whatever that may be) must at all costs change. Our way of life must be protected from the threat of the new, from wherever that may come. Another gentler version is in the cosmology of aboriginal peoples, whose sacralised concept of their home territory saw it as the location of a cycle of generational renewal and rebirth, which excluded the idea of development in any unprecedented and unwarranted directions.

Massey has expressed some sympathy for forms of resistance to the overpowering pressures from global market forces which find expression in spatially defined identities – 'our place', its community, its heritage, its natural environment – though it is to be noted that such imaginative constructions usually embody a temporal history too. The idealised object is 'our place and its distinctive past'. But she does not in the end believe that such largely spatialised responses to the forces of imposed one-dimensional change can suffice in sustaining productive conceptions of political agency. Resistances of place are not enough to hold off the power of global markets.

Doreen Massey's fundamental argument is therefore neither exclusively for space (despite the crucial categorical importance of space to the field of geography) nor for time. It is rather for space *and* time conceived essentially as related to one another, with each neither excluded nor diminished by the other. Social and historical analysis, she eloquently argues, should be constructed in terms of space-time relations. Massey is arguing that both wholly spatial and wholly temporal modes of analysis entail what in another terminology could be called reification. That is to say, they treat social entities or processes as thing-like and one-dimensional, whereas they need to be understood to be both in actuality and potentiality as processual, and multidimensional. Massey's argument is for the recognition of contingency and openness in social and political processes, as the essential root of ethical and political progress.

Her theoretical and political positions in this respect draw substantially on the arguments of Ernesto Laclau and Chantal Mouffe (1985), who have theorised the democratic political sphere – in opposition to the essentialism and determinism of the Marxist tradition – as a properly deliberative and agonistic one, 'without guarantees', as others have put it. Laclau and Mouffe argued that Marxism, even in both of its most sophisticated Althusserian and Gramscian forms, remained dogmatic in its assumptions, ascribing unity to social formations, and necessary goals and intentions to class actors, which have no empirical basis. Instead, they urge the primacy of the political and the contingent, proposing that the distinctive contribution of both Lenin's and Gramsci's conceptions of agency was that they did acknowledge the necessary role of political action in situations whose outcomes could never be fully determined.[2] They argue that classes and other political subjects are not the pre-existing ground for political action, but in part have to be constituted through it.

Mouffe (2000) went on to argue for a conception of 'agonistic democracy', urging the importance of the prosecution of conflicts of fundamental value (against the silencing contemporary ideas of political consensus), though nevertheless framed within a commitment to the democratic process itself ('adversaries, but not enemies' is her formulation of this). Massey has been influenced by this conception of political action, and the necessity to create alliances and movements through establishing 'equivalences' between claims by different collective subjects, and she argues for a spirited conception of political action and mobilisation, in varieties of situation. She has brought together her idea of spatially constituted differences between subject positions with a commitment, which draws on Mouffe and Laclau's work, to an anti-essentialist democratic politics (Massey 1995a, 2005). Her underlying commitment is to spatial and human connectedness in relation to these various struggles. The description by Ehrenreich and Hochschild (2003) of the exploitative global pattern of arrangements for childcare which compels Filipino mothers to care for the children of affluent Californian mothers, while their own children have to be left in the care of grandparents and aunts, is an example of the kind of spatial patterning of social relations which can be recognised through this critical framework. In some instances, the specific political causes which Massey embraces can seem somewhat disparate and insubstantial, yet many flourishing initiatives – such as some of those she describes in part three of *World City* (2007) – indicate that this politics has real roots.

There is a 'postmodern' cast to Massey's positions, following from her hostility to what she sees as the authoritarian transformative ambitions of different versions of modernism. Her ontology is somewhat constructivist, in her view that subjects are constituted through discourse and action, and her conception of agency is somewhat pragmatist. Notwithstanding this, her substantive political and ethical commitments have remained unyielding.[3]

The attention given in Doreen Massey's work – especially her *For Space* (2005) – to the categories within which geographical and political thought takes place follows from the necessity at certain times to interrogate philosophical categories themselves. This is notably when dominant conceptualisations cease to be able to capture phenomena that call for description and explanation. Massey's argument has been with what she believes to be misleading frames of thought, which produce false ways of accounting for and imagining human experiences. She has been engaged in an argument against what can best be described as ideologies, as framing preconceptions whose consequences both for thought and politics she regards as harmful.

Both of the extreme poles of the antinomy between spatial and temporal modes are imaginable modes of thinking, and indeed each has some contemporary sway. To the mainly temporal mode belongs an active, teleological, transformative politics of modernisation. To the largely spatialised mode belong various kinds of conservative defence of what *is*, in this spatially located 'here and now'. Like other ideologies, each of these has been and remains capable of ensuring its own condition of reproduction. Ideologies which deny the existence of relevant differences within spaces may well be linked with instruments which diminish or remove them from political consideration in practice. For example, the Soviet project of modernisation attempted to impose uniform models of economic and political organisation wherever its power allowed.[4] The 'Washington consensus' was similarly intolerant of societal differences from its own neoliberal perspective. Conversely, ideologies which deny the existence of teleologies, or indeed possibilities of change, can be associated with political practices which repress these. Authoritarian theocracies, whether dominated by Catholicism or Islam, have had this character. Ideologies – even systems of abstract presuppositions like those of space and time – are thus engines which drive both substantive ways of thinking and political practices. Even the most abstract categories of thought are supported by material powers, and in their turn give them legitimacy. It is thus valuable that Massey has given so much attention to the meanings and implications of such categories.

What is being advanced through Doreen Massey's categories of space-time relations is in effect a conception of how she would like social relations and politics to be. What connect places and spaces to one another in her view are the social actions of those individuals and collectives who inhabit and move between them in time. 'Time-space compression' is the consequence of a vast expansion of human and technological powers of many kinds. Massey has sought to map the consequences of this situation, in terms of those differentially affected by these powers. She has explored how access to these powers is distributed, for example by class and by gender, and how these distributions can be contested. She develops a view of interconnectedness which she sees as giving rise to potentials for liberation and democracy, but locates these in the contexts of the social forces and structures – the 'power geometries', as she puts it – which oppose these. Her own view of 'time-space compression' is in fact a fairly positive one, since processes of globalisation, which to some are mainly sources of disorientation and dissolution, seem to her to contain new possibilities for the construction of improved ways of life. The connections between places, and also their temporalities (the historically produced ways of life in which communities live), though often exploitative or oppressive, can also bring the possibility of change.

I suggested earlier that the discipline of geography had been able to understand the new world dominated by global market forces better than most sociologists had been able to do. One reason was that geographers took to the globalised world as their natural (if new) intellectual habitat, while most sociologists were still caught up in nation-specific fields of study. The apparently abstract focus on changing relations of space and time made it possible to recognise the significance of many more phenomena – the growth of mega-cities, the functions of shanty towns as zones of transition within them, the development of 'virtual' circuits of financial exchange, 'mobilities' of many kinds – than was possible in the conventional lexicon of much sociology and political science. Changing relations of time and space are constitutive features of this emerging social order, and it is for this reason that the focus on them by geographers and others has been so productive of understanding.

But it emerges that in Massey's work the primary object of study has all along been essentially social, and indeed political. Reflection on the categories of space and time has made possible new understandings of the actions of the human beings[5] located within them, as a consequence of the changing relations of time and space. Massey's contention is that connectedness and relatedness between places and spaces, and between societies and cultures with their specific histories or 'times', has increased, one would say as a consequence of expanded human and technological powers.

Relational Approaches in Psychoanalysis

A link can be made between the relational theory that was being developed in the context of the discipline of geography, and the 'object-relations theory' which in this period has become the most influential framework in psychoanalysis in Britain. Certainly the idea of relations and the relational is very central to both ways of thinking. Doreen and I have been close colleagues on the journal *Soundings* for more than 15 years, and agree politically about far more than we disagree, yet we have previously found few ways of bringing together the two fields of geography and psychoanalysis which are respectively central to our interests. So it seems worth exploring these possible affinities a little further.

One notes to begin with that psychoanalysis and associated forms of thinking have also become more 'relational' in recent decades than they previously were. Freud's original framing of psychoanalysis was somewhat individualist, with the self and its desires placed in primordial conflict and competition with the desires of others. But later developments in psychoanalysis, for example in the writings of Klein and Winnicott, gave a greater emphasis to the idea that infants were related to parental figures (often referred to technically as 'objects' because of their unconscious internal representations) from birth. The narcissism that had been declared to be a primordial condition by Freud came to be seen as a response to a failure in a primary relationship, not its natural precursor (Rustin and Rustin, 2010). Another earlier split, between a predominantly 'internalist' view of personality development which tended to ignore actual relationships in families and beyond, and perspectives which emphasised the effects of the 'real' environment, has diminished, as many psychotherapeutic practitioners have come to recognise that 'internal' personality formations, and the actual relationships within which persons are located from birth, need to be thought about together. Furthermore, in different psychoanalytic traditions, such as relational psychoanalysis in the United States (Mitchell and Aron, 1999) and the post-Kleinian tendency in the United Kingdom, much attention is now given to interactions between analyst and client in a therapeutic relationship in which the analyst is no longer imagined to be an unmoved observer and interpreter of what is happening. As our wider society (in Britain and the USA at any rate) has become more individualist in the last three decades, the psychoanalytic and psychotherapeutic fields have become more attentive to social connectedness and dependency.

One question is whether any particular vision of society and social relationship follows from a psychoanalytic perspective whose primary spatial location is the clinical consulting room, and whose primary focus of interest is on the inner world of individuals, usually within a context in which intimate 'family' relationships, past and present, are of most concern. One can ask if any specific politics follows from this perspective, and if so, whether it is anywhere specified what this might be. But even if the answer to this question is a qualified negative, one can ask if conceptions of the social good may nevertheless be implicit in this work, potentially capable of translation into a wider-than-clinical sphere.

The 'relational' dimension of this perspective emerges in its presence in various spheres of public policy, often as a minority voice. For example, in devising forms of care for the old, including those suffering from dementia, it is being argued that the quality of human attention accorded to subjects with diminishing capacities is essential to their well-being, since it sustains their sense of being through relation to another (Davenhill, 2007; Dartington, 2010). Or, in responding to the extreme difficulties in parenting of babies which sometimes occur (perhaps as an ultimate consequence of wider social injuries and deprivations, and attempted escape from these through drug or alcohol abuse), it is argued that it is the parents' capacity to tolerate the needs of their babies, and the emotions aroused by them, which is the crucial issue. The question is whether, if the parents themselves can arrive at a sense of being held-in-mind, they may recover the capacity to hold-in-mind their infants.

Or, in considering the difficulties that sometimes occur in situations of adoption, it has been argued that it is helpful to realise that accompanying a troubled relationship between adoptive parents and child there is likely to be another 'internal' relationship-in-the-mind between adopted child and birth parent, or adoptive parent and wished-for birth child. There may be a multitude of 'subjects' in the situation, some of them denied or unrecognised (Rustin, 1999).

Or, in encountering many situations of difficulty in families, at any age, it has been found that past relationships – even going back a generation or two – may have a continuing shaping hold on the present, sometimes providing an unconscious template by which current relationships are unknowingly shaped.

Or, in considering the stresses on workers and professionals in many areas of the education, health and welfare systems, it is noted that their primary relationships of choice, with their students, pupils, patients or clients, and with their field of practice – their teaching research, their professional interest and expertise – may both feel under attack. Such attacks come from demands to follow bureaucratic procedures, to achieve 'outputs', to succeed in competitions (e.g., positions in league tables), which may be quite insensitive to the primary needs of the work in question (Rustin and Rustin, 2010).

Or, in thinking about how members of organisations can work productively together, and how managements can facilitate this, attention is given to whether the anxieties inherent in an organisation's work can be recognised, thought about, 'contained' in the mind of those who hold responsib, so that attention can be given to the tasks of an organisation. Different organisational 'missions' are liable to carry their own specific anxieties. Those working in psychiatric settings will be preoccupied, consciously and unconsciously, with the perils of mental illness for all who come into contact with it; those in schools with the difficulties of learning, with the risks of failure, and with the problems of maintaining commitment to the work of learning when the risk and fear of failure is always present. Those in prisons have to cope with a fear of the potential resistance or violence of those whom they are imprisoning, and perhaps also with a despairing knowledge of their inability to do much to help many of those in their charge. All these are relational states of mind, of one kind or another.

The underlying ontology of a relational psychoanalysis approach includes the following elements. One is the idea that individuals are always located within a context, in which relation to others, both in actuality and in unconscious fantasy, is central to their being. Donald Winnicott memorably wrote, 'there is no such thing as a baby, only a mother and baby' (1966), but one can broaden this assertion to say that all individuals depend on others to constitute their being. Thus there is no teacher unless there are also pupils, no actor unless there is an audience to watch them perform, no doctor or nurse unless there are patients. The reason why retirement from work can be a crisis is because it may take from the self both a real and an imagined context of interaction, of shared being, which may not be capable of being adequately replaced.

A second element is the idea that individuals are multiple subjects, both in their actual present, and in their remembering and self-imagining (what will also, be in psychoanalytic terms, their unconscious.) One psychoanalytic discourse thinks of this in terms of 'parts of the self'. Joyce McDougall, in her *Theatres of the Mind* (1986), writes about conscious and unconscious 'scripts', the problem for human actors being to avoid being overwhelmed by the power that unconscious scripts may have to take over the conscious self. Although in psychoanalytic clinical practice, the issue may often be to clarify the relation between 'infantile' and more grown-up parts of the self, this idea of multiple selves is not simply a developmental or 'temporal' idea, but also refers to coinciding – we might say 'spatial' – differences between different latent subjects within the self.

A third idea is that the capacity to reflect, to hold in mind the different and contradictory aspects of those 'objects' which the self is related to, is critical for mental integration and development. Under emotional stress, these different aspects become split off from one another, and 'binary' ways of thinking ensue. A world of the good 'inside' and the bad 'outside', or the good 'us' and the bad 'them', is constructed. We are all familiar with endless versions of such binary oppositions and antagonisms, of racial, religious, ideological, territorial and many other kinds. These are termed 'paranoid-schizoid' modes of mental functioning in the Kleinian psychoanalytic context, in contrast to 'depressive' ones which take account of the complexity and capacity to suffer of the other. The idea that the attainment of a capacity for reflection, especially in emotionally charged situations, is the crucial index of maturity has become crucial to contemporary psychoanalysis, largely through the contribution of Wilfred Bion, influencing many of its component theoretical tendencies.

A great deal could be said theoretically about the importance of the toleration of differences as an attribute of mind, within the contemporary psychoanalytic theory of the personality. Freud argued that the unconscious knows boundaries of neither time nor space. A more philosophically developed formulation of this idea is to be found in the writings of the Argentine psychoanalyst Jorge Ahumada, who has made use of Russell's theory of logical types to explain the modes of unconscious mental functioning.[6] Ahumada (1991) argues that the unconscious thinks, so to speak, in classes, from whose universal qualities and attributions the particulars of experience are not able to be differentiated. In healthy mental life, symbols are used to name and discriminate different attributes of reality, and to identify connections between them, whether by analogy or as cause and effect. Where the mind is unduly dominated by states of unconscious fantasy, internally generated symbolic representations become fused with perceptual reality, and distort and restrict its apprehension.

Doreen Massey has not generally been concerned with extreme situations in her consideration of the consequences of the denial of difference. But they do occur. Entire societies have at times been invaded by states of mind in which it was claimed that all differences from an absolute norm must be eradicated. The persecution of heresy, or the cause of racial purification, are examples of the expunging of differences which are 'spatial' in nature – that is to say, their intolerance is of different modes of being in the present. Struggles to destroy a class enemy (one remembers that 'liquidate' was a term in common use), whether bourgeois under Stalinism or aristocratic under the Jacobin Terror, defined their intolerable field of difference in historicist terms, as obstructions to the march of progress. These states of mind at certain times acquired a virtually psychotic character, in which it became impossible to differentiate between fantasy and reality. However, these theoretical arguments regarding the reality principle in psychoanalysis, essential for the apprehension and toleration of differences, cannot be further pursued here.

Space and Time in Psychoanalysis

More can be said about the significance of the dimensions of space and time in psychoanalysis. One starting point is Roger Money-Kyrle's (1971) idea that the fundamental challenge in psychological development is to come to terms with the 'fact of life' – that is to say, the realities of generation and gender, of the human capacities that arrive but are then lost with time, and of the different capacities that are given and denied by gender. This generalises Freud's theory of the Oedipus complex, and the disappointments of infantile desires which this makes inevitable. Generation is a temporal concept and gender is spatial, in a metaphorical sense.

The psychoanalytic process aims to make possible, within a chosen temporal and spatial boundary (the set time and space allotted to analytic sessions), an exploration of psychic space and time, with the purpose of making possible a fuller engagement with and acceptance of differences of many kinds. One can say this is an exploration in both time and space: time in so far as the experiences of the past – in fantasy and in actuality – are made accessible to reflection and transformation in the 'here and now' of the transference relationship; and space in so far as differences and indeed conflicts can be negotiated, both internally and externally in this setting. The analytic process is intended to make possible a widening and deepening of emotional and mental life, a freeing of thought and feeling from closures of various kinds. The boundaries set by the frame are themselves usually the focus of much analytic work, since resentment and denial of them may reveal the structure of unconscious desires and anxieties. This is true for Lacan's technique of deliberately 'breaking the frame' by unexpectedly ending an analytic session, as well as for the 'English style' in which these boundaries are firmly adhered to.

A zero degree of psychological closure to the experience of space-time relations is found in autistic states in children. Such children often cannot bear any separation from their objects of attachment (which would imply difference), nor change of any kind other than the most repetitious or mechanical. I once heard an account of psychotherapeutic work with such an autistic boy patient, who partially recovered. This boy had chosen to live in a completely frozen world, in which continuities neither of space nor time were recognised. He had no effective memory or sense of anticipation, nor was there any preparedness to connect things happening in one place with things happening in another. When the boy started to recover, the means by which he came to recognise the properties of time and space was by discovering his own powers of causal agency – by realising that he could make a difference. He found, for example, that blowing out the candle was what made it go dark. It was the development of a sense of self, of being an active agent in the world, that made differences of time and space begin to be tolerable rather than experienced as threats of annihilation. Here was a little boy learning the lessons of Heraclitus and Parmenides about the way things both change and persist in time. There are, no doubt, more parallels to be explored between the space-time relations of Doreen Massey's geography and space-time relations as they are conceived in psychoanalysis.

Space-Time Relations in Geography and Relational Approaches in Psychoanalysis

What, if anything, does a psychoanalytically informed 'relational' perspective have in common with the relational dimensions of Doreen Massey's writing? Both perspectives give great emphasis to relations with others, in effect to desirable engagements with difference. In the psychoanalytic sphere, this relation is mainly conceived in reparative and inclusionary terms. How can all these aspects of experience (in an internal world, in a family, in an institution, even in a larger society) be given their due recognition, be accepted for what they are, be made objects of thought rather than expelled from consciousness and turned into objects of hatred? 'How can the sense of reality be defended from internal attack?' is a continuing question, since it is recognised that individuals and groups are liable to defend themselves through rejection of reality itself, through various forms of denial and unthinking.

In Massey's thinking, the relational focus is often more combative and conflictual. Differences need to be recognised and accepted, but the unequal and unjust geometries of power need to be addressed through political and social action. I don't think this contradicts the imperatives of a psychoanalytic way of understanding differences and relationships, but it is distinct from it. The explanation of this difference is itself situational, a matter of spatial location, one might say. The psychotherapists find themselves primarily engaged in the work of emotional repair. Even when they are working with institutions, they are usually set the task of making them work better, more inclusively, more thoughtfully and kindly, not of attacking them for their deficiencies, still less attempting to sweep them out of the way. Doreen Massey often writes as a political citizen and activist, offering descriptions and explanations of injustices, showing how they are connected to larger systems of power relations. In this role she nevertheless proposes a mode of political action which is reflective, committed to understanding, respectful of differences, responsible, while still being prepared to identify opponents and seek to defeat them.

Both conceptions are essentially relational, in their different ways. One could say that each carries with it its innate risk. For the reparative psychoanalytic approach, the risk is of excessive compliance with existing institutions and powers. Therapists and consultants can usually only work within the social spaces which the dominant order accords to them – although because this order (in most places) is not a monolith, and because it may harbour various potentialities for development, these spaces can sometimes be significant. For the pursuit of an agonistic politics, the contrary risk is that in the heat of conflict, 'differences' may once again become collapsed into binary antagonisms, equivalences become flattened into identities, and tolerance of complexity and the capacity for free thought greatly diminished.

What none of us knows is what the emergence of relational perspectives like these, of different but perhaps nevertheless compatible kinds, may now socially signify. An earlier 'reparative' moment in psychoanalysis in Britain was connected historically to the broader movement for social reconstruction which followed the experience of the Great Depression and the Second World War, and made its small contribution to the development of the welfare state. Might these developments in theory and practice in geography and in psychoanalysis perhaps now indicate the scope for a broader development of relational social thinking, at the end of a period in which neoliberal individualism has been so much the dominant way of thinking?

Acknowledgements

I am grateful to Ash Amin, Margaret Rustin, Joe Painter and David Featherstone for their valuable advice during the preparation of this paper.

Notes

1   It was only in a semi-jocular spirit that Doreen Massey and the other founding editors of *Soundings* contemplated calling it *The Kilburn Times*, before we realised that while Kilburn might signify a rather unfashionable neighbourhood to us its residents, to readers elsewhere the metropolitan designation of London might signify more.
2   Rustin (1988) set out a critique of Laclau and Mouffe's *Hegemony and Socialist Strategy*.
3   Richard Rorty sustained in his writing a similar conjunction of a constructivist philosophy and an outspokenly socialist ethics and politics (Rorty, 1999).
4   Tony Judt's (2007) [2005] *Postwar: A History of Europe since 1945*, pp. 165-196, describes the effects of Soviet rule over most of Eastern Europe after 1945 in these terms.
5   Bruno Latour (2004), and perhaps Massey too, would say not only human beings, but other kinds of actant too, such as the elements of the environment itself.
6   Among the sources for these ideas are the Chilean psychoanalyst Matte Blanco (1975) and the anthropologist, psychiatrist and systems theorist Gregory Bateson (1973).

Chapter Five
Space, Democracy and Difference:
For a Post-colonial Perspective

David Slater

Introduction

For Massey (2005: 11), a thorough spatialisation of social theory and
political thinking can 'force into the imagination a fuller recognition of the
simultaneous coexistence of others with their own trajectories and their
own stories to tell'. Perhaps nowhere is this statement more pertinent than
in the annals of democratic theory, within which the dominant Western
orientation draws a veil over its own particularity, whilst generalising for the
world. In developing an alternative view on the spatiality of democracy,
I want to discuss three interwoven topics: (i) the imperial context within
which this alternative approach needs to be located; (ii) a consideration of
key features of the democratic imperative; and (iii) the illuminating relevance
of a post-colonial perspective on democratic theory.

The Imperial Context

Some time ago Edward Said (1993: 66), when discussing culture and
imperialism, made the point that whereas in much recent cultural theory
the problem of representation is deemed to be central, nevertheless it is
rarely put into its full political context, a context which is primarily imperial.
Whilst it is certainly the case that since the mid-1990s more research has

Spatial Politics: Essays for Doreen Massey, First Edition.
Edited by David Featherstone and Joe Painter.
© 2013 John Wiley & Sons, Ltd. Published 2013 by John Wiley & Sons, Ltd.

been done on the interconnections between cultural studies and imperial power, the durability of imperial power continues to exert a pivotal influence on a whole range of theoretical and political questions, not least the way we interpret the meanings, trajectories and relationalities of democracy.

When – in, for example, the context of the United States – we begin to raise questions about the meaning of imperialism or imperial power, there is a frequent tendency to become fixated on the economic. However, as the Retort (2005) group has argued, the circumstances that oblige the state to act in the way it has are rarely straightforwardly 'economic'; rather it is the interweaving of compulsions – economic, spectacular and geopolitical – that reveal the 'American empire's true character'. What is surely critical here is the working out of the interaction between the agents of power and the nature of the interweaving of compulsions or issues of geopolitical direction. A traditional Marxist account would tend to emphasise the point that the agents of state power would take decisions that benefit the interests of capital, since the imperial state expresses those interests. But is this enough?

The notion that imperialism has to be related to the 'driving force of the economic factor', in the phrase used by Hobson (1988 [1902]: 74), or to what Arendt (1979: 125) suggested was 'the central political idea of imperialism', namely 'economic expansion', also shows that the concentration on the economic is not confined to the Marxist persuasion.[1] It seems to me that although we cannot abandon the importance of the economic as one of our compulsions of spatial expansion – here we can mention the continuing search for resources and raw materials (Klare, 2002) and new markets as well as cheap labour – still they are not enough to explain the imperial drive. Moreover, does the concept of 'compulsion' offer us sufficient analytical breadth? Can we encapsulate the factor of 'mission',[2] or more broadly 'desire', into compulsion? Was and is 'manifest destiny' or a notion of geopolitical predestination a compulsion or rather an expression of one kind of political desire?

We have arrived here at a rather fundamental and complex question: how, in the specific context of the United States, do we account for the imperial mentality? William Appleman Williams, in his books on empire and the history of American diplomacy (see, for example, Williams, 2007), emphasised the importance of a certain 'conception of the world' that fluctuated between a primarily economic vision and a largely cultural and political orientation. This imperial conception of the world was traced back to the end of the eighteenth century/early nineteenth century. What is particularly relevant about Williams' view was that he underlined the significance of the emergence of a discourse of imperial power, as reflected, for instance, in the Jeffersonian concept of an 'Empire of Liberty'. During the nineteenth century and beyond, the idea that the United States was a nation with an imperial destiny was expressed and articulated by a host of writers, politicians and business leaders (see Slater, 2010: 190).

Throughout the nineteenth century and on into the twentieth, the emerging imperial power of the United States was codified through the Monroe Doctrine of 1823, the concept of 'manifest destiny', which surfaced in the 1840s at the time of the US–Mexican war, the 'Open Door policy' of the 1890s and the Roosevelt Corollary of 1904. The Monroe Doctrine was a statement of differentiation from Europe, and of leadership of the Americas, with the newly emerging republics of Latin America being referred to as the 'southern brethren'. The notion of manifest destiny encapsulated the sense in which the United States was envisaged as possessing a geopolitically predestined role of becoming a global power, and the Open Door policy was designed to open the world's – and in particular China's – market to American-manufactured goods. The Roosevelt Corollary assigned to the United States a role of 'international police power'; hence wherever there was considered to be chaos and disorder, or threats to civilisation in the countries to the south of the United States, intervention would be justified.

A key element of this emerging imperial power was the development of an invasive desire which can be seen reflected in the fact that between 1798 and 1895, well before the birth of the Soviet Union, there were 103 US interventions in the affairs of other countries (instances of the use of US armed forces abroad), ranging geographically from China and Japan in the East to Argentina and Nicaragua in the West (Zinn, 1996). By the eve of the Spanish–American War, an editorial in *The Washington Post* declared that 'a new consciousness seems to have come upon us – the consciousness of strength – and with it a new appetite, the yearning to show our strength .... Ambition, interest, land hunger, pride, the mere joy of fighting ... we are animated by a new sensation. We are face to face with a strange destiny. The taste of Empire is in the mouth of the people even as the taste of blood in the jungle' (Zinn, 1996: 290).

This ethos of invasiveness goes well beyond the drive of economic expansion. There is a sense of a multifaceted desire to expand and penetrate which is already deeply rooted in the society and which has geopolitical, cultural, psychological, economic and military aspects. Equally, it needs to be taken into account that an overarching sense of pre-eminence carried with it a subordinating attitude to other peoples and cultures. There were two variants of this subordinating mode of representation, and we can find them both expressed at the beginning of the twentieth century. The second variant has endured longer than the first, even though the first variant has certainly not disappeared.

First, one has a belligerent, openly racist discourse as expressed by Theodore Roosevelt in 1902. At the time of the colonial war in the Philippines, Roosevelt observed that the conflict signified 'the triumph of civilisation over forces which stand for the black chaos of savagery and barbarism'. Furthermore, he went on, 'the warfare that has extended the boundaries of civilisation at the expense of barbarism and savagery has been for centuries one of the most potent factors in the progress of humanity' (quoted in Kramer, 2006: 169).

Second, and in the same year, President-to-be Woodrow Wilson argued that 'we must govern ... and they must obey as those who are in tutelage'. He continued, 'they are children and we are men in these matters of government and justice' (Wilson, 1902: 13). This second variant expresses a condescending, ethnocentric attitude which is deeply rooted in US (and Western) society and which has affected and continues to influence the chemistry of US–Third World relations. It is a constitutive component of imperial discourse and Schoultz (1999), for example, gives us many pertinent examples of the continuity of this ethnocentric attitude in the history of US–Latin American encounters.[3]

The spatial expansionism associated with the imperial drive is frequently explained, as intimated above, in the context of the intrinsic and penetrative logic of capitalism, and the role of a civilising mission is seen as a pretext for underlying economic objectives, such as the pillage of resources and raw materials. Interestingly, some writers, such as Caroline Elkins (2005) in her painstaking study of the twilight of British colonialism in Kenya, stress the importance of the 'civilising mission', not as a superficial pretext for foundational economic motives, but as a central objective of the imperial/colonial project itself. Elkins writes that 'Britain's far-flung empire was united by a single imperial ethos, the "civilising mission" …. The British were to bring light to the Dark Continent by transforming the so-called natives into "progressive citizens" ... they were self-appointed trustees for the hapless "natives" who had not yet reached a point on the evolutionary scale to develop or make responsible decisions on their own' (Elkins, 2005: 5).[4]

Hence, it can be suggested that the desire, *inter alia*, to 'civilise', to provide tutelage (or, more recently, to democratise), and to maintain order and security have been and remain objectives that have permeated the imperial project. And they are objectives which have been presented as thoroughly justifiable. When we connect desire to objectives it is crucial to highlight the significance of political will. Desire is not enough; it has to be canalised into an effective and sustainable will, which takes us into the ambit of state policy and intervention. I would argue that it is not capitalist interests per se but rather governmental power which expresses and mediates the expansionist desires of the society as a whole; it is a site or point for the condensation of imperial desire.[5]

Clearly, an imperial project requires a continuing build-up of capacities – military, economic, technological, counter-insurgency and geopolitical – and it is evident that throughout the twentieth century and beyond, the United States has constructed a formidable war machine capable of intervening anywhere on the planet. In fact, the Department of Defense in 2006 states that what is now necessary is the 'ability to surge quickly to trouble spots across the globe' (Department of Defense, 2006: v). In addition, according to Johnson (2004: 24), Defense Department figures show that by the early part of the twenty-first century the United States had 725 overseas military bases, bases which for Johnson are a sign of militarism, the 'inescapable companion of imperialism' (ibid.).

Moreover, in the economic sphere the United States accounts for approximately a quarter of the world's GDP (in 2007) and has a higher figure than the next three most important economies (Japan, Germany and China) combined (see Callinicos, 2009). In terms of industries of the future, for example nanotechnology, a full 85 per cent of venture capital investments in this area went to US companies, and a similar picture emerges with biotechnology (Zakaria, 2008). Therefore, it can be argued that US imperial power has the military, economic and technological capacity to reproduce itself, even without considering the importance of 'soft power' or the effectiveness of its cultural projections. It is in this context that we need to pose the question of legitimacy and the role therein of democracy, or more specifically 'democracy for export'.

Democracy and the Geopolitics of Difference

There is a quite pervasive narrative that encourages us to believe that the West, and specifically the United States, is a dutiful carrier of democracy to other parts of the world (Smith, 1994). 'Making the world safe for democracy' was a key leitmotif of the Woodrow Wilson years, and towards the end of the First World War Wilson asserted that America had always stood for the principle 'that no nation should seek to extend its polity over any other nation or people, but that every people should be left free to determine its own polity, its own way of development, unhindered, unthreatened, unafraid, the little along with the great and powerful' (quoted in Keane, 2009: 361). This interpretation of foreign policy expresses a deeply rooted sense of idealism and innocence whereby the reality of US imperial power is concealed behind a veil of posited benevolence and respect for the rights of others. In actual fact, it can be suggested that the export of democracy has been an integral part of America's sense of mission; it is the way the nation has justified to itself, as well as to others, the universality of its principles. Further, within this narrative, a world of governments more like the United States would engender a safer and more prosperous world and hence it could be asserted that interventions to introduce democracy to foreign polities would be entirely justifiable on both security and ethical grounds. As Steel (1995: 18) reminds us, the United States is virtually alone in the world in declaring the promulgation of democracy to be a major foreign policy objective.[6]

In order to propel the argument further, I would like to make the following five points on democratic politics, points that connect with both the imperial

context and the following discussion of post-coloniality and the geopolitcs
of knowledge:

1   The democratic is a classic example of a polysemic term that is dependent
on the different discourses that give the term its meaning. Concepts such
as, for example, 'popular democracy', 'imperial democracy', 'indigenous
democracy' and 'liberal democracy' reflect the continuing attempt to
ground a definition of democracy that will always be contested. What
needs to be underlined here is that it is a vision of 'liberal democracy' or
'market-based democracy' that has become dominant in an era of
neoliberal globalisation, so that what is a specific form of democratic rule
has come to be traditionally seen as the only sensible form democracy
can take. In contradistinction to the conventional belief in plurality,
democracy under neoliberalism has come to assume a uniform character,
linked into the symptomatic notion that 'there is no alternative'. But, as
Barber (2004: 177) pungently reminds us, 'the neoliberal ideology of
privatisation that has dominated political thinking in the last several
decades, and that has been the unspoken context for the American
approach to "democratising" the globe, has in fact had a corrosive effect
on democratic governance'. Market fundamentalism has done little for
democracy, and it 'disdains democratic regulation with dogmatic
conviction' (Barber, 2004: 177).
2   Following on from the above, much of the contemporary discussion
surrounding the need for democracy has been characterised by belief in
the desirability of exporting the Western and especially the US model of
liberal democracy for non-Western societies.[7] There is a governing
assumption that Western democracy has a universal validity acting as an
already available template that non-Western polities need to follow. This
presumption neglects the fact that democracy must be seen as a diverse
phenomenon which can only be sustainable if it emerges from indigenous
roots; what is required is a struggle for what may be termed 'demo-
diversity'. And it is here that we can make a link with Massey's concern
for multiplicity and difference. So, whilst it can be argued that space is
the sphere of multiplicity, a 'realm of multiple trajectories' (Massey,
2005: 89), this insight can be extended into the field of imaginations,
understandings and theorisations. In this sense, the theorisation and
understanding of 'demo-diversity' can be nothing but a multiple exercise
that can enrich our discussions of space and democratisation and avoid
the straitjacket of universal, unilateralist thought.
3   Democracy, as long as it is to remain vibrant, requires an underlying
process of democratisation in the sense of the renovation of the forms of
participation and the development of autonomy, as reflected in the will
and capacity of citizens to be reflexive and critical of governmental
authority. With the extension of democratic principles into civil society

as well as the democratisation of the institutions of the state, one can envisage a process of 'double democratisation', where there is a mutually sustaining interaction between two connected processes. Moreover, this process of double democratisation needs to be situated in a spatial context whereby the struggle to decentralise power to local and regional levels can embody the territorial amplification of democratic politics. This is very much an open process since there is no guarantee that the local or the regional levels of a polity are going to be naturally more democratic than the central level; it will all depend on the nature and direction of the social struggles at the various levels. This point is similarly expressed by Massey (2005: 166) in her discussion of space and place, where she argues that openness and closure should not be expressed in abstract spatial forms, but in the terms of the social relations through which the spaces are constructed. Or, in a related passage, it is proposed that the differential placing of local struggles within the 'complex power-geometry of spatial relations is a key element in the formation of their political identities and politics' (Massey, 2005: 183).

4   Democracy can be seen as the attempt to organise political space around the commonality of society with efforts to constitute a unity of one people, or to ground popular sovereignty. Conversely, democracy has also been conceived of as an extension of the logic of equality to broader spheres of social relations – socio-economic equality, racial equality, gender equality, territorial equality and so on – so that in this case democracy involves respect for differences. Thus, as Laclau (2001: 4) puts it, the 'ambiguity of democracy' can be formulated as requiring unity but only being thinkable through diversity. In addition, it is worthwhile adding here that, following Mouffe (2000: 2–3), the specificity of modern democracy comes from the contingent articulation between two different traditions: on one side there is the 'liberal tradition constituted by the rule of law, the defence of human rights and the respect of individual liberty, and on the other the democratic tradition whose main ideas are those of equality, identity between governing and governed and popular sovereignty'.[8] In addition to what Laclau refers to as the 'ambiguity of democracy', one needs to remember that for societies of the periphery or global South, the development of democracy has always been affected by the invasive logic of Western interventions, and this factor has frequently been crucial, as was the case, for example, in Guatemala in 1954, Iran in 1953, Chile in 1973 and Nicaragua in the 1980s.

5   In relation to globalisation and the geopolitics of difference, the attempt to export and implant one vision of democracy as a unifying project across national frontiers, as in the Iraqi and Afghan examples, not only clashes with the logic of differences alluded to above, but comes up against nationalist and indigenous resistance. To be liberated and democratised at the point of a gun hardly constitutes an effective way of spreading and

justifying any model of political rule. Also, the dissonance between the model of Western liberal democracy and other indigenous, community-based forms of democracy, perhaps most visibly evident in the Bolivian case (see for example Hylton and Thomson, 2007), is rooted in two opposed logics: one, originating outside, associated with an imperial mentality, the other an authocthonous, inner-produced logic of democratic politics. Contested combinations of the two, giving us different types of 'hybrid democracy', seem likely to become more prevalent, but how these types will evolve is difficult to predict, much depending on the orientation and strength of social movements.

It is here that we might refer to a general point made by Nietzsche, in discussing the uses and disadvantages of history for life. Nietzsche (1990 [1874]: 62) was concerned to highlight the negative impact of giving too much weight to the past; what was needed was the setting free of the 'plastic power' of men, people and cultures. By plastic power, Nietzsche was referring to the 'capacity to develop out of oneself in one's own way, ... to heal wounds to replace what has been lost, to recreate broken moulds'. Confidence in the future, he continued, depends on being able to forget at the right time and to remember at the right time. The danger, he suggested, was that an 'excess of history', when individuals or societies are weighed down by the burden of the past, can attack life's plastic powers, so that the past can no longer be employed as a 'nourishing food' (ibid.: 120). It is precisely in this sense of a 'confidence in the future', where the past is used as a nourishing food and not regarded as a permanent burden, that social movements such as the Zapatistas in Mexico have found a way to chart a new future whilst using elements of the past, such as the connection with Emiliano Zapata. Similarly, related movements in Bolivia (as seen in the mobilisations against water privatisations and gas pipeline investments) and the Landless Rural Workers' Movement (MST) in Brazil have been able to develop a plasticity of thought and action which has kept them vibrant and sustainable.[9]

For a Post-colonial Perspective on Democratic Politics

In the opening quotation to this chapter, Massey talks about the simultaneous coexistence of others and the need to listen to the stories of others. A more spatialised vision on democracy and international relations would certainly take this suggestion into account. However, as Gruffydd Jones (2006: 12) reminds us, in the introduction to his edited collection on decolonising international relations, 'one of the major characteristics of the imperial imagination of IR (international relations) and social inquiry more generally is the simple but often massive failure to mention or remember events, processes and scholarship of the non-Western world'. What we encounter in the social science literature in general is a strong tendency to erase the importance of the non-Western world, and this tendency is both deeply rooted and manifest in the contemporary period.[10] This erasure or deletion can be associated with a related tendency to implicitly frame the non-West in a negative light.[11]

A post-colonial take on these issues might develop a critique of Westocentric interpretations of democracy by looking at three interrelated questions: (i) the limits of democracy in First World countries; (ii) the real nature of West–non-West connections; and (iii) the complexities of democratic politics in peripheral societies.

A first point that needs to be made concerns the notion that it has always been in the core Western countries that democracy has been more fully developed. Markoff (1999), in his analysis of where and when democracy was invented, examines the writing of constitutions, competitive electoral parties, representative institutions, accountability, secret ballots and the extension of suffrage. He concludes this survey by noting that in the past two centuries the great innovations in the invention of democratic institutions have generally not taken place in the world's centres of wealth and power. In other words, the picture is more complex than is conventionally assumed.

Taking a comparative view, and looking at the extension of voting rights for women as a key element of democratic progress, it can be seen that the West has not always been in the vanguard. For example, whilst a number of Latin American countries gave women the vote before the1940s (Ecuador in 1929, Chile in 1931, Brazil and Uruguay in 1932 and Cuba in 1934), certain First World countries were considerably later in extending voting rights to women (France in 1944, Italy in 1946, Japan in 1947, Belgium in 1948 and Switzerland in 1971).

Turning to issues of race, in the United States the African American population faced varied formal obstacles to the exercise of their voting rights including problems of literacy, arbitrary 'character' requirements and the threat of violence against those who turned up to vote. It was only in 1965 with the Voting Rights Act that African Americans were able to acquire their full voting rights, although even in the 2000 election irregularities with respect to race were encountered in the state of Florida.

In addition to the uneven extension of voting rights and the restricting role of a racial hierarchy, there are other limiting factors concerning democracy in the First World. The corrosive influence of organised crime, corruption, the role of restricted access to financial resources for electoral purposes and growing voter apathy all ought to lead us to sound a critical and cautionary note when the so-called superiority of Western democracy is routinely taken for granted. Moreover, it is important to distance ourselves from a view that equates the West with democracy and leaves out of account the Western history of fascism. The dark side of the West in the context of totalitarian rule needs to be kept in mind when broad generalisations are made about the posited superiority of Western democracy.

Furthermore, post 9/11, the 'war on terror' has led to a notable erosion of democracy in the United States, as well as elsewhere in the West. The organisation Human Rights has documented the continual circumvention of law in the treatment of prisoners and detainees in Afghanistan, Guantánamo and in Abu Ghraib. In a connected analysis of imperial democracy, Eisenstein (2008) stresses the importance of the remilitarising of American society. She notes how a culture of pre-emptive strikes and unilateral power plays out on both the battlefield and everyday life in the United States. In addition, it is argued that 'war is our cultural metaphor. We war on drugs, on AIDS, on cancer, on poverty, on terrorism ... war is a danger to democracy because it justifies and therefore normalises secrecy, deception, surveillance and killing'. As part of the remilitarisation of life, hierarchy, surveillance, authoritarianism and deference become a way people live both inside and outside the centres of military power (Eisenstein, 2008: 28–29).

Further to the contentious presupposition that the West and especially the United States constitutes *the* template for global democracy, there is also the pervasive notion that the West has been responsible for the promotion and diffusion of democracy to the non-Western world. However, in actual fact, and as intimated above, it can be shown that the West, and the United States in particular, has acted as a *terminator* of Third World democratic governments. For instance, in 1954 in Guatemala, a CIA-backed coup overthrew the elected government of Arbenz, who had initiated a programme of land reform which was strongly opposed by the United Fruit Company. Once in power, the Guatemalan military reversed the Arbenz land reform, and in the period from 1960 to 1996 more than 200,000 people from the largely indigenous population were killed (Grandin, 2004). A similar intervention took place in Iran in 1953 where the democratic government of Dr Mossadegh was overthrown in a CIA-backed coup. In Latin America related interventions took place in 1965 in the Dominican Republic and in Chile in 1973, where a reforming democratic government was violently terminated by a US-backed Pinochet dictatorship. In the case of Nicaragua, the Sandinista government, which had comfortably won an election in 1984, was destabilised by the Reagan administration and subsequently lost the 1990 elections.[12]

The termination of independent democratic governments has had its reverse side – namely a record of support for pro-Western dictatorships. In South America, military regimes in Argentina, Brazil, Chile and Uruguay were not destabilised but rather supported (see Slater, 2007). A similar pattern has been evident in other regions of the global South, and in the current era support for undemocratic regimes such as those in Saudi Arabia, Egypt and Uzbekistan contradicts the notion of a Western diffusion of democratic politics. Western, and especially US, backing for non-democratic regimes has a long history and it is symptomatic of mainstream writing that such support is rarely mentioned.

As far as the past and present of democracies in the global South are concerned, one is confronted by a blanket of Western prejudice. Societies of the periphery have been characterised as being 'pre-democratic', or in need of democratic guidance, or in transition to democracy, or as 'emerging democracies' and so on. Amartya Sen (2007, 2010) is particularly pertinent at this juncture, since in a short section on the 'global roots of democracy' a number of insightful points are made.

First, Sen notes that although modern concepts of democracy and public reasoning owe much to European and American analyses and experiences (for example from the European Enlightenment), nevertheless to extrapolate backwards from these comparatively recent experiences in order to construct a long-run dichotomy between the West and non-West would be to make a very odd history. Looking at the continually posited Greek origin of democracy, Sen suggests that there is a great reluctance to take note of the Greek intellectual links with other ancient civilisations to the east and south of Greece, in spite of the fact that the Greeks themselves showed a vital interest in talking to ancient Iranians or Indians or Egyptians. For Sen, while Athens was the pioneer in getting balloting started, there were many regional governments which went that way in the following centuries. For example, some of the contemporary cities of Asia incorporated elements of democracy in municipal governance, such as the city of Susa in south-west Iran which for several centuries had an elected council, a popular assembly and magistrates who were elected by the assembly. Equally, it needs to be noted that, as Sen suggests, democracy is not just about ballots and votes; it is also about 'government by discussion', and here there is a long history across the world, as Sen emphasises (2007: 53). A similar argument is made by Ayers (2006), who reminds the reader of indigenous forms of democracy in an African context.

On the basis of the examples that Sen (2007: 54) brings together, he makes the valid suggestion that the 'Western world has no proprietary right over democratic ideas'. Such ideas belong to the world – or, expressed in a Zapatista language, these kinds of ideas belong to a world in which many worlds fit.

Overall, and linking back to previous comments on social movements, it can be proposed that it is in the global South where the democratisation of democracy or the emergence of a counter-hegemonic democratic politics is particularly evident. In other words, far from being a passive recipient of a diffusing Western democratic template, it is in the countries of the South that alternative forms of the deepening of democratic politics are to be found. What is at issue here is alternatives to the hegemonic version of democracy, to a formalised neoliberal democracy that stresses the market, elections, competition between political parties and an acceptance of the given disposition of power relations. The alternatives, or what we might term 'demo-diversity', include a vibrant emphasis on participation in democratic politics whereby across the different worlds of the global South social movements have played a crucial role in bringing a reinvigorated democratic imagination to civil society.[13]

Concluding Remarks

In the global context of a resurgence of imperial politics, the promotion and diffusion of one particular interpretation of democracy acts as a potentially effective legitimisation of a reasserted form of the penetration of Third World sovereignty. The call for democracy is a powerful one since it evokes a movement towards equality, progress and a modern form of political engagement and rule. Who could be sensibly against the spread of democracy and freedom? But the key issue lies in the orientation and content of democratic politics. For example, the injunction to democratise creates an asymmetry between those issuing the injunction and those subjected to it, or, in other words, between those who 'democratise' and those who are being 'democratised'. In the official Western – or more specifically US – template, the parameters and effects of the imperial perspective tend to be occluded. The geopolitics of the 'seer' are normalised and naturalised so that alternatives are excluded or rendered inappropriate. In the case of democracy, this means that the enabling potential of learning across cultural divides is negated, and in its place there is a tendency to prefer processes of imposition based on the supposition that the Western template is of universal validity.

Alternatively, and this would be a central component of a post-colonial approach, instead of the unilateral export of the Western model of liberal democracy to the rest of the world, what is needed is the creation of a space or multiple spaces in which learning about the different cultures of democracy can take place in a spirit of mutual respect and recognition. In a sense, this might be called the democratisation of the study of democracy, which can be associated with a globalisation of the study of globalisation. For democracy to flourish, it has to be home grown and autonomously sustained, not implanted from outside as part of a justification of a subordinating imperial project. At the interface of the imperial and the democratic there are a series of antagonistic tensions that can never be resolved since crucially the imperial ethos violates the foundational and dialogic roots of the democratic spirit. If that spirit is to be protected and sustained, the imperial mentality has to be continually challenged and superseded so that democracy may flourish in an open and creative manner.[14]

Notes

1   A more nuanced vision is to be found in the more recent work of David Harvey (2003), where it is argued that imperial power can be looked at in terms of two distinct but intertwined logics – a territorial logic (a political project on the part of actors whose power is based in command of a territory and a capacity to mobilise its human and natural resources towards political, economic and military

ends) and a capital logic – both coming into play to account for accumulation by dispossession.

2   The concept of mission goes back to the nineteenth century in US history, but it has remained an important trope throughout the twentieth century. In the 1950s, at the peak of the Cold War, Secretary of State John Foster Dulles talked of the 'incompleted mission of the West', when referring to the expansion of freedom (Department of State 1971: 23).

3   A contemporary example of a continuing tutelage relation can be found in the United States Southern Command document for 2008, which is entitled 'Partnership for the Americas', and which deals with US–Latin American relations. The document specifies the need for US 'training of Latin American countries' in the area of internal security, whereby the United States ought to help the Latin American countries 'better understand the linkages of their entire governmental apparatus, as it pertains to their internal security, sovereignty and cooperation' (USSOUTHCOM, 2008: 14).

4   The Zanzibari novelist Abdulrazak Gurnah takes up this theme in relation to colonialism in East Africa and pens the following sardonic passage: 'The Empire selflessly brought us knowledge and education and civilization and the good things that Europe had learned to make for itself and which until today we have still not learned to make for ourselves. Instead of being left in our degenerate darkness for centuries to come, within a few decades we were opened up and dragged into the human community' (Gurnah, 1997: 73).

5   For a related but contrasting approach which falls within a Marxist frame, see, for example, Panitch (2000), who interprets the imperial state in the context of the unfolding dynamic of capitalist development.

6   For a detailed analysis of US foreign policy in the context of democracy promotion, see Barber (2004).

7   For some further examination of this theme see Slater (2009).

8   As Mouffe (2000: 6) explains it, it is important to distinguish the discourse of political liberalism from today's 'neoliberalism', the latter being interpreted as follows: 'neo-liberal dogmas about the unviolable [sic] rights of property, the all-encompassing virtues of the market and the dangers of interfering with its logics constitute nowadays the "common sense" in liberal-democratic societies ...'.

9   For a detailed review and analysis of Latin American social movements, see Stahler-Sholk, Vanden and Kuecker (2008).

10 For example, in a recent book on culture and international relations, Lebow (2010), in a text of more than 750 pages, excludes the Third World from consideration and there is no discussion of North–South relations.

11 For example, Dunn (1993), in his well-known text on Western political theory, writes that as a resource for understanding the political history of the world, 'western political theory has great strength and no effective surviving rival' (Dunn, 1993: 130). In other words, the non-Western world is relegated to a non-existent status as regards understanding of the world's political history; it has no theoretical value.

12 For some detail on these examples, see Slater (2007). Although these cases of US intervention are well known, it is always necessary to remind ourselves of

their importance, especially since the official discourse of democracy promotion consistently ignores the historical record.

13 For instance, considering India, Appadurai (2007) connects a new politics of hope to an explosion of civil society movements which make some use of the conventional practices of democracy such as open legislative deliberation and a vigorous sense of the accountability of rulers to ruled, but which also have generated a new range of practices that allow poor people to exercise their imaginations for participation – practices that include techniques for self-education and ways of pressuring state and party officials to act on basic needs without falling into machine patronage and vote-bank politics.

14 The last two to three pages closely follow my argument in Slater (2009).

Part Two
Regions, Labour and Uneven Development

Part Two
Regions, Labour and Uneven Development

Chapter Six
Spatial Divisions and Regional Assemblages

Allan Cochrane

Introduction

The regional question continues to haunt geography and, equally important, to frame a whole set of public policy initiatives and political debates. In the context of her wider engagement with, and creative reformulation of, geographical thinking, Doreen Massey has not only helped to challenge conventional ways of thinking about regions but has also opened up new ways of understanding what they might be and how they should be understood in political as well as economic terms.

In simplistic terms, she is sometimes understood to have developed a relational approach but this understates the significance of her contribution. At an early stage, in ways that remain remarkably prescient, she fundamentally questioned the extent to which it makes sense to think of a 'regional problem'; she has set out an agenda that highlights an active process of uneven development through spatial divisions of labour; she has emphasised the need to 'rethink' the region in ways that explore how social relations stretched across space also come together to make up regions; she has critically engaged with notions of a global city region to set out London's role as a node of political and economic power.

Here a review of some of Doreen's contributions to the literature provides the basis for an argument that indicates how her approach can be mobilised and taken further in analysing the ways in which regions are constructed

Spatial Politics: Essays for Doreen Massey, First Edition.
Edited by David Featherstone and Joe Painter.
© 2013 John Wiley & Sons, Ltd. Published 2013 by John Wiley & Sons, Ltd.

politically. The aim is to explore changing forms of regional governance and power relations and their significance. Rethinking the region offers us a way of moving beyond simple territorial conceptions, recognising the continuing processes by which regions are made and remade and confirming the need to understand the way in which apparently 'regional' processes are themselves the product of relations that stretch across space in ways that reflect and reinforce uneven development, helping to generate forms of political responsibility which cannot be contained within the region.

Questioning the Assumptions

In Britain the discussion of 'regions' has historically been lodged within a policy discourse in which it is almost impossible to use the word 'region' without adding the word 'problem'. From the identification of the 'distressed', 'depressed' or 'special areas' in the 1930s (Ministry of Labour, 1934; Commissioner for the Special Areas, 1935, 1936; Hannington, 1937) to the contemporary rhetoric of renewal and regeneration, it remains difficult to escape the notion that it is in the 'regions' that change is needed – England's South-East may be a region in the language of geographers (even if they, too, tend to focus on the more familiar spaces of the North-East, the North-West and South Wales – namely the old industrial regions) but it is rarely thought of as a 'region' in any other sense.

And, of course, the identification of a 'regional' problem also implies that the solution is to be found in the areas facing the problems (although even in the 1930s there were proposals to limit growth in the South-East to benefit the 'regions'). The policy geography implied (and often stated explicitly) is one that requires the 'regions' to be more productive, to become competitive and – above all – to move on from the failed industries of the past, and – by implication – the failed political and trade union traditions of the past.

From the start, this version of geography – this political version of economic geography – was one that Doreen Massey questioned and challenged. One important paper did so head on, posing the question 'in what sense a regional problem?', and she pursued the arguments forcefully in a series of publications (in particular; Massey, 1979; 1995b [1984]; Massey and Meegan, 1982; Allen and Massey, 1988). Massey's development of the notion of spatial divisions of labour subverted any understanding that simply drew on explanations in terms of industrial sectors (i.e., blaming the 'regions' for having the old industries etc.). Instead she highlighted the mechanisms that generated and (equally important) utilised uneven development and inequality to generate profits, as part of the normal operation of decision making in a capitalist economy.

The arguments being developed here are important in their own right. They question taken-for-granted beliefs and highlight the lazy assumptions of politicians, policy makers, academics and others who focus attention on the people being forced to adjust to new economic realities, rather than on those benefiting from them. In that sense, of course, these are not merely 'academic' arguments. They are politically significant because they highlight the extent to which the old ways of thinking, the dominant policy nostrums, simply missed the point. And at the same time, they set out an agenda for economic geography that enables it to move beyond the charting of inequality and division to emphasise the need to explore and recognise the mechanisms that generate that inequality and those divisions. Rather than explaining it all in terms of market failure or market imperfection, Massey draws attention not only to the extent to which but also – more important – the ways in which the spatial divisions of labour that are inherent in the working of Britain's particular form of capitalist economy foster continued uneven development and inequality.

She identifies a complex set of spatial negotiations through which regional hierarchies may be reproduced, reworked or reimagined, and highlights the ways in which regional differences are actively managed through the investment strategies of major corporations. Instead of being more or less taken-for-granted spaces across which social and economic forces work their magic, this way of thinking about regions makes them active participants in the shaping of social and economic change. And, of course, these insights are not just relevant to the 'regional problem'.

They also point towards the possibility of what later came to be identified as 'relational' approaches. At first glance these conclusions may seem rather uncontroversial, saying no more than that it is necessary to understand that there are no geographical absolutes since space is necessarily defined in the relations between actors, objects and context. But once one begins to take these points seriously then it is the ways in which these relations work out that matter as much as the patterns that emerge.

Conceptualising Spaces and Places

This becomes apparent in Doreen Massey's discussion of place. At the core of this discussion of what makes places places is an apparent paradox or conundrum, namely that 'places seem to become both more similar and yet lacking in internal coherence; home-grown specificity is invaded – it seems that you can sense the simultaneous presence of everywhere in the place where you are standing' (Massey, 1994b: 162). This insight into the complexity of everyday life is, of course, not the end of the matter, merely the beginning of a process of thinking. Massey argues that rather than

> thinking of places as areas with boundaries around, they can be imagined as
> articulated moments in networks of social relations and understandings, but

where a large proportion of those relations, experiences and understandings
are constructed on a far larger scale that what we happen to define for that
moment as the place itself, whether that be a street, locality, or a region.
(Massey, 1991b: 28)

From this perspective place is defined through the coming together of
a whole series of overlapping relationships and understandings, each
juxtaposed to the other. Massey argues that places can effectively be seen as
'a momentary co-existence of trajectories and relations; a configuration of
multiplicities of histories all in the process of being made' (Massey, 2000a:
229; see also Massey, 2005). Focusing on place as 'meeting place' in this
way is an important counter to arguments that see the local (local places or
even communities) as more or less pre-given entities, products of tradition,
just waiting to be disrupted by the forces of globalisation. On the contrary,
it suggests that they give meaning to the global through the way in which
globalisation works out and is constructed in practice. From this perspective
globalisation must always be placed.

At the same time, the emphasis on a 'momentary' articulation of a
multiplicity of intersecting relations, understandings and trajectories
highlights the need to think about places as dynamic, porous and continually
being made, never completed. In this sense, in other words, space has to be
understood 'as an open and ongoing production' (Massey, 2005: 55).
Massey draws attention to the 'throwntogetherness' of place, defined through
the almost accidental juxtaposition of communities, groups and individuals
linked out into wider networks yet actively – locally – engaged together in the
making up of neighbourhoods, cities and regions (Massey, 2005).

In broad terms, the conceptualisation of regions implied by these ways of
thinking can be stated relatively simply. Regions are 'a series of open,
discontinuous spaces constituted by the social relationships which stretch
across them in a variety of ways' (Allen et al., 1998: 5). This way of thinking
is enormously liberating because it makes it possible to move beyond the
search for some sort of underlying 'region' as more or less fixed entity, or
something which has the potential to become fixed, defined through
necessary sets of relationships. Instead it emphasises the need to explore the
ways in which regions are actively constructed, made and remade, or
assembled through political as much as economic or social practices.

The notion of assemblage is one that has found a range of expression, in
the hands of a series of theorists: from Deleuze and Guattari (1987) to Ong
and Collier (2005); from Sassen (2006) to Latour (2005) and Li (2007).
And, of course, the specific understandings vary between them. However, in
this context, Aihwa Ong's summary is helpful in stressing that, 'As a field of
inquiry, assemblage stresses not structural hierarchy but an oblique point of
entry into the asymmetrical unfolding of emerging milieus ... the promiscuous
entanglements of global and local logics crystallise different conditions of

possibility. This conceptual openness to unexpected outcomes of disparate political and ethical intentions suggests that outcomes cannot be determined in advance' (Ong, 2007: 5). In strong echoes of Doreen Massey's approach, this suggests that rather than starting from some pre-existing or logically coherent framework into which everything fits, it is necessary to consider the way in which meanings (and projects) are constructed through negotiated practices between agents of one sort and another.

Re-thinking London and the South-East

In work that I and others undertook with Doreen in the 1990s, the focus deliberately shifted from the 'regions' as problems to the South-East of England, a region that was explicitly and historically defined through success. We argued that in the 1980s and 1990s the 'South-East' was best interpreted as a 'growth' region and specifically a region of neoliberal growth. We sought to explore and analyse the South-East of England as a neoliberal heartland, the product of overlapping social, political and economic relations, which stretched across space in ways that showed little or no respect for the regional boundaries imposed upon them.

This was, in other words, an expression of a continuing critique of the approach that sees the UK's regions as the problem – on the contrary, the focus was directed towards the contemporary South-East as the problem, or, at any rate, the spatial conjuncture that had to be explored and understood. It is apparent that the interpretation of the 'regional' problem as one which defines the poorer regions in terms of the failure of their traditional industries or their (consequent) dependence on state subsidy of one sort is not only unhelpful but actively misleading, and generative of a politics that is incapable of challenging the powerful or the causes of inequality.

In the Thatcherite discourse of the 1980s, the South-East was often presented as a model of deregulated growth – the positive example of what could happen if only the constraints of bureaucratic red tape and wasteful government spending could be lifted to free up the initiative of entrepreneurs. But, in practice, such formulations understated the full extent of some of the hidden subsidies, for example as reflected in changes to the tax system that favoured higher-paid employees – even 'deregulation' delivered gains to specific sectors, such as financial services and the City, both directly and through the windfall gains of managing large-scale privatisations. And this was accompanied by supposedly place-neutral infrastructure investment in major initiatives, from a whole series of bypasses to the building of the M25 as well as investment in Heathrow airport. More recently, New Labour in power gave us something called the sustainable communities plan (ODPM, 2003) which promised growth across England, but in practice sponsored property development and infrastructure investment in a series of so-called 'growth areas' clustered in the Greater South-East, alongside further major investment in the Thames Gateway and, still more recently, infrastructure and mega projects associated with the 2012 Olympics. Increasingly ambitious targets were set for the building of housing that would ensure that the regional economy would not be undermined by significant labour shortages or upward pressure on wages and salaries (see Barker, 2004, for a discussion of the perceived need for housing growth).

In her work with Ash Amin and Nigel Thrift (Amin, Massey and Thrift, 2003), and later in her analysis of London as a world city (Massey, 2007), Massey sets out a view that emphasises the way in which London's position in a regional hierarchy and as a definer of national economic priorities has largely negative effects on the wider political economy of the UK. London's economic success (in terms of the narrow definitions of GVA [gross value added] and average income) is based on an economic model that works not only to reinforce regional inequality, but also to undermine any prospect of more generalised economic growth. London and the South-East acts as a magnet sucking in highly skilled labour, while also being dependent on the existence of a low-paid, casualised workforce at the bottom end of the labour market (this can be contrasted with the more positive conclusions drawn by Gordon, Travers and Whitehead, 2008, who present it as a major source of national economic benefit).

Effectively this also turns the claims sometimes made for the South-East as a 'growth region' that drives the British economy on their head. Instead it is suggested that the economic logic that drives London and the South-East (the Greater South-East) works as a drain on the rest of the country. Amin, Massey and Thrift (2003) highlight the extent to which graduates are drawn to the region from elsewhere in the UK, in ways that reinforce the centralisation of the knowledge economy in London and the South-East, at the expense of other parts of the country. If the South-East is a 'core' region, then, according to this analysis, it is one whose relationship with the rest of the British economy is rather peculiar (and potentially dysfunctional), since – as Allen et al. (1998) note of the 1980s and 1990s – by sucking in talent, public and private investment, 'the political economy of the south east as "growth region" operated in ways which restricted the possibility of growth in other regions of the UK' (Allen et al., 1998: 119).

But this is a political as much as an economic process. Rather than accepting that Britain's North/South divide is a consequence (however distorted) of the working out of market forces or of London's status as world city, Amin, Massey and Thrift (2003) argue that the centralisation of power in London and the South-East means that a 'significant element of national policy making effectively functions as an unacknowledged regional policy for the South Eastern part of England' (Amin, Massey and Thrift, 2003: 17; see also Marvin, Harding and Robson, 2006). They suggest that national economic policy is overly influenced by the state of the regional economy in

London and the South-East, with steps being taken to restrain the economy when the region is 'overheating', even when the rest of the country still has significant capacity for growth. Conventional approaches to 'regional policy' (like those favoured by Labour governments of the 1960s and 1970s) were concerned to find ways of equalising development by shifting investment from the South-East and other growth regions (at the time including the West Midlands) to economically weaker regions, and debates around regional policy have tended to focus on this approach and the extent to which it delivered what was promised (see, for example, Moore and Rhodes as far back as 1973). This argument turns such debates around by focusing on regional policy in practice, and highlighting its spatial implications.

This is a South-East that can be identified topographically – it describes a space that can be (and has been) mapped in various ways by geographers and others in recent years (whether as the Greater South-East, a polycentric urban mega region, a set of administrative government arrangements or a metropolitan core) (cf. Gordon, 2003; Dorling and Thomas, 2004; SEEDA, 2006; Pain, 2008). But it is also a South-East that can ultimately only be understood in topological terms – that is, through the ways in which it is linked and connected, pulling in and reaching out to, sets of social relations. In Allen *et al.* (1998), an attempt is made to capture this through the metaphor of the doily, which lays out its lacy template across the region, connecting up places of growth, but also (to mix metaphors) somehow stretching out tendrils to areas (apparently far away) whose everyday lives can be understood as connected to (or part of) the social phenomenon that is the 'South-East'.

In this context, Massey's discussion of London's role as a world city is particularly suggestive, because of the way in which it refuses to accept any overall logic of globalisation, instead placing it and problematising it (Massey, 2007). She highlights the extent to which London – even in the case of financial services – is in practice part of a UK market for most of the goods and services in which it trades. As a result she argues that the taken-for-granted emphasis on London's global role, as reflected in policy documents and the political practices of national as well as Greater London governance, is better understood as a (neoliberal) political strategy than as any 'objective' economic reality.

Alongside this she suggests that the 'geographical concentration' of the very wealthy in London and the South-East transforms it 'into a self referential echo chamber [that] reinforces their distance from the rest of us' (Massey, 2007: 66), carrying with it a powerful set of discourses clustered around notions of 'competitive individualism and personal self-reliance' (Massey, 2007: 38–40). Her powerful critique spatialises the politics of the contemporary UK to the extent that the apparently local politics of London, nevertheless predicated on a notion of its 'global' role, are also part of a process of national political manoeuvring in which the 'Reinvigoration of London … represents the rise of a new elite, and the culture in which it is

embedded' (Massey, 2007: 49). If this reinforces the argument about
London's position within the UK – and its role as generator of inequality –
it also has specifically local consequences, in reproducing poverty and
inequality within the city. As Massey puts it, 'London's poor ... and those
without higher level skills, are caught in the cross-fire of the city's reinvention'
(Massey, 2007: 64).

But, at the same time, the global aspects of London as 'world' city are
significant, too, because of the way in which they make it possible to consider
what Massey (2004) identifies as global 'geographies of responsibility'.
While acknowledging the force of the argument of those who point to the
cosmopolitan nature of London's population (and even the lived experience
of everyday multiculture in the city), she points to some of the underpinning
reasons why London has taken on this role. She questions the easy
assumptions of those (such as Ken Livingstone, despite her sympathetic
engagement with him in debate and discussion) who simply celebrate
London as the home of diverse populations drawn in from across the globe
and apparently generating 'vibrant' spaces of interaction and cross-cultural
learning. Instead she notes that one of the reasons why London has such a
high proportion of migrants and of black and minority ethnic people in its
population is precisely because the operations of those in the City of London
have helped to force them out of their own countries, reducing opportunities
there while generating unequal growth in London (Massey, 2011b). In
other words, therefore, she argues that it is necessary to think about ways of
generating a local politics that somehow takes account of the extent to
which (and even the possibility of) celebrating the daily practices of the
global in the local depends on sacrifices made far away.

What she reveals is an unequal power-geometry of interdependencies.
This goes beyond the recognition that particular places can only be
understood in relation to each other – through webs of interconnection
and dependency. Politically what matters for her is that the places that gain
from these relationships owe a debt to those that are disadvantaged by it.
This is an insight that can be mobilised at all sorts of geographical scales
and between all sorts of places (between the neighbourhoods of London,
between the regions of the UK, between the country and the city and
between the cities of the world and so on). In terms of a global sense of
responsibility, it draws attention to the ways in which the relationship
between London and the cities of the global South works to benefit London
at their expense and calls for a positive engagement to challenge those ways
of working (see, for example, the discussion of what a radical political
agenda for London might look like in Massey 2004, 2007 and 2011b, and
Massey and Livingstone, 2007).

But this is emphatically not just a politics of charity based around
identifying the undoubted injustices faced by the poor of the world, or by
those who are somehow affected by London's impact on their lives. It is a

politics based on the understanding that these people are themselves part of London – their activities are already part of the space that defines London and gives the city its identity. Many of London's residents are themselves directly connected (through chains of family, friendship and culture) to this wider set of people, but that understates the extent of the linkages, of the interdependence, which helps to give meaning to the notion of London as a world city. The spatial politics of London – its power-geometries – combine and entangle global, national, regional and local sets of relationships.

What Is a Region?

All this takes us a long way from traditional ways of understanding regions. In this world, regions are always in the process of being made, never finalised. And they are constituted out of relations that stretch across the boundaries given by the administrative map-makers – who themselves are frequently challenged to develop new ones as those boundaries shift, as reflected, for example, in the relatively mundane remapping of the South-East's so-called 'growth areas' in the sustainable communities plan (ODPM, 2003), which constructed new border areas stretching across several government regions (most dramatically in the Thames Gateway, but also, for example, in Milton Keynes and the South Midlands, which stretched from Luton in the South up to Corby in the North). What this highlights is the need to focus on the relations rather than the maps – and never to forget that these are relations of inequality: the making up of regions, however it is done, continues to be underpinned by and reflect processes of uneven development still there. As Massey reminds us, the apparent success (or at any rate prosperity) of London and the South-East cannot be disconnected from the apparent 'failure' of other regions – indeed its very social and economic dominance helps to define them as 'failures' or 'problems' (cf. Leunig, Swaffield and Hartwich, 2008, who argue that the solution is for people to move to London and the South-East).

It is still possible (and indeed often helpful) to work with a territorial understanding of regional spaces. 'Relational' thinking does not mean the end of territory, but rather reinforces the need to identify how territories are made up, constructed or assembled. In other words, they cannot be taken for granted; nor can it be assumed that just because the name remains the same, the 'region' is the same. But it is nevertheless still possible to identify regions, just as they may be the focus of policy discussion or even political (and sometimes popular) identity. They may have fuzzy edges and they may be defined through the wider sets of connections that come together in them, but still recognisable. What matters, however, is always to recognise and acknowledge that they are actively constructed – they are always in the process of becoming, and it is this that needs to be the focus of attention.

Like other territorial formations, this means that the construction of regions is fundamentally a political process. Of course, this does not mean that regions may be produced simply by some feat of political imagination (or, indeed, simply by the stroke of a civil servant's pen). On the contrary, even if Massey's approach can be contrasted with David Harvey's search for 'structured coherence' (Harvey, 1985), it is necessary to acknowledge the extent to which the construction of identifiable regions (and regions with which people identify) relies on the existence of particular forms of economic and social relations (even if they are in contention and even if they cannot be fully contained within the borders by which they appear to be defined).

To return for a moment to the South-East, the growth region dynamic model of the 1990s and early twenty-first century was always in tension with another version of the story – one that started from the privileges of suburbia. This version (itself also deeply rooted in the social and economic relations of place) effectively denied the regional story and instead stressed the priorities of residents who wanted to protect their existing lifestyles and amenity without being threatened by the downside of growth. This has a long history (see Hall *et al.*, 1973, who emphasised the ways in which the planning system focused on the 'containment of urban England' and operated to protect the property values of existing residents of London and the South-East and to disadvantage those seeking what has come to be called 'affordable' housing). But it has more recently been strongly espoused by the politicians of the Home Counties, with the added spice of the language of environmentalism and sustainability (cf. Foley, 2004; Commission on Sustainable Development in the South East, 2005).

Following the election of the coalition (Conservative/Liberal Democrat) government in the UK in May 2010, this understanding seems to have gained a new respectability. There has been a move away from the setting of ambitious regional housing targets, which required local authorities to plan for significant increases in supply, mainly delivered through the private sector (at what by 2008 had become quite unrealistic levels). At the same time, there has been a move away from the identification of growth areas, and away from the 'local delivery vehicles' initially tasked with delivering new housing on a significant scale – most of these special-purpose bodies have been translated into more modest, locally based partnerships, or else (even in the case of London Thames Gateway) responsibility is being given to local councils. More significantly, perhaps, it is local government that is expected to decide on levels of house building in their areas, without the imposition of any nationally determined priorities.

In the South-East, in other words, the suburban councils of the Home Counties are once more being empowered to hold back development that undermines their taken-for-granted spaces of privilege (see, e.g., Charlesworth and Cochrane, 1994; Cochrane, 2011). The extent to which this vision can be sustained over any length of time is, of course, questionable.

Today, the weakness of the housing market is an expression of the financial crisis and economic recession. But any return to (unequal) growth will simply revive the pressures in the region, and even without growth London and the South-East continues to experience a housing market which seems unable adequately to supply the needs of its population. In a sense, the new regional arrangements, which have also brought an end to regional development agencies, promise a return to 'business as usual' in which uneven development is reproduced by the apparently 'hidden' hand which Doreen Massey has so consistently sought to uncover. In institutional terms, it looks as if the regions of the North may keep a version of their existing regional bureaucracies in some rebranded form, while in the South-East what are to be called 'local enterprise partnerships' will mean something quite different.

Conclusion: Relational Thinking and Regional Assemblages

What Massey gives us is the ability to explore these shifts. Instead of assuming that there is some sort of objective truth to the regional nomenclature handed down from the map-makers of the civil service or elsewhere, her work allows us to approach the region as a rich set of contestations and interactions. In this context, a suburban 'region' is no less a 'region' than (what used to be called) a planning region. There is no 'proper' region against which the South-East can be judged. In writing about any set of socio-spatial arrangements (community, neighbourhood, city, region etc.) there is always a danger of reverting to some sort of ideal type or idealised form to which all must aspire and few (none) attain. Instead it is necessary to focus on the active processes by which a range (a multiplicity) of different actors and agencies are assembled to make up regions, which may themselves only ever be provisional, but may nevertheless also operate as identifiable forms of political and economic territory.

Tracing the lineage of particular ways of thinking about space and spatiality, cities and regions, is always risky. Those from whom one draws inspiration or on whose ideas one seeks to build are not always convinced by the ways in which their ideas are reinterpreted or reworked. So it is with some trepidation that I have enrolled Doreen Massey's work into the development of an approach to the politics of regions which draws on the notion of assemblage. But it seems to me that the active processes of nego-tiation implied by that notion and the extent to which, as Ong (2007: 5) puts it, it assumes 'the promiscuous entanglements of global and local logics', legitimises such an enrolment.

Returning to the 'regional question' or even 'regional problem', thinking in this way makes it possible to build on Massey's insights. Politically, as I (and John Allen) have argued elsewhere, it highlights the extent to which, 'the governance of regions, and its spatiality, now works through a looser, more negotiable, set of political arrangements that take their shape from the networks of relations that stretch across and beyond given regional boundaries. The agencies, the partnerships, the political intermediaries, and the associations and connections that bring them together, increasingly form "regional" spatial assemblages that are not exclusively regional, but bring together elements of central, regional and local institutions', as well as agents identified with the public, private and not-for-profit sectors (Allen and Cochrane, 2007: 1163).

But it also provides a reminder that the form of those assemblages also has to be understood as provisional, even temporary, which is a valuable message at a time when the institutional infrastructure of New Labour's regional architecture is being dismantled so thoroughly. Of course this does not mean that there are no longer 'regions', but it does mean that it is still more important actively to explore the contingent ways in which they are put together, being reborn and reimagined under different spatial headings (from city regions to polycentric mega regions).

Equally importantly, it reinforces the need to recognise that place is more complex than any straightforward topographical mapping might suggest. The apparently distant may be very tightly bound up in the ways in which particular places have to be understood. And some of the activities that appear to be close by – next door – may in significant ways be less closely connected. It is this understanding – so important to Massey – that makes it possible to call for a politics of responsibility stretching across space which is not simply an altruistic call for the strong to support the weak, but, rather, which recognises the mutual dependence of those who may at first sight seem fundamentally separate. From this perspective, distance can no longer simply be measured in terms of miles or kilometres (or scale in terms of hierarchical level), which means that the politics of regions (and the politics of cities) can no longer be limited to the administrative boundaries they have been allocated.

Acknowledgement

Thanks to Juan Arredondo and John Allen for talking to me about these things.

Chapter Seven
Making Space for Labour

Jamie Peck

Introduction

This chapter reflects on Doreen Massey's role as a (proto) labour geographer, suggesting that some of her early contributions articulated a distinctive position on the problematic of labour geography, even before the appellation, the field and the project existed as such. Some of this was apparently lost in the autocritique of economic-geographical practice that effectively launched labour geography proper, as a normatively agency-centric project, in the late 1990s. Rereading labour geography through the lens of the 'industrial restructuring' approach from the first half of the 1980s, and in particular through Massey's (1984) seminal *Spatial Divisions of Labour*, need not be an exercise in intellectual or political nostalgia. It can also shed new light on the restructuring present, while perhaps providing some pointers to some of labour geography's futures. As debates in the field of labour geography have recently turned to the challenges of placing and positioning labour's agency, rather than simply prioritising it per se, so there may be new things to learn from the time (and the place) where these questions were originally opened up – not least in Doreen Massey's work at the beginning of the 1980s.

The project of 'labour geography' is conventionally dated to the 1990s – a productive current in the wider field of pluralising, 'new' economic geographies, which in some ways it has since transcended altogether. In the wake of Andrew Herod's (1997) initiating statement, new lines of research have

Spatial Politics: Essays for Doreen Massey, First Edition.
Edited by David Featherstone and Joe Painter.
© 2013 John Wiley & Sons, Ltd. Published 2013 by John Wiley & Sons, Ltd.

flourished, inter alia, on the (new) politics of organised labour and on social (or community) unionism; on cultures of work and the social dynamics of employment change; on the geographies of labour markets and labour processes; and on a wide range of working-class struggles and campaigns.[1] One of the dominant motifs here has been that of labour as an active geographical agent, as a space-maker and scale-shaper engaged in the construction of its *own* spatial fixes (though hardly under conditions of its own choosing). Characteristically, this work is consciously positioned not only after the market-centric economic geographies of neoclassical location theory – which reduced labour to an input factor – but also after those capital-centric economic geographies of more orthodox Marxist theory, which dealt for the most part with abstract, disembodied labour in accounts that privileged the dynamics of capital accumulation.

Spatial Divisions of Labour is typically, though somewhat awkwardly, bracketed with this latter tradition, along with a handful of similarly foundational texts which effectively shaped the Marxist canon in critical human geography (see, especially, Harvey, 1982; Smith, 1984; Lefebvre, 1991 [1974]). But while Massey's book – in political terms, certainly – was very much a product of both its time and its place, its pioneering form of heterodox political economy was *simultaneously* a reflection of, and a step beyond, prevailing analytical and methodological currents in radical economic geography. In one sense, the book marks an historic inflection point, between the failure of the British Labour Party's modernisation project and the rise of Thatcherism; in another, it operationalises ways of seeing and engaging that transcend this conjuncture altogether. Just as Julie Graham would remark, on reviewing the second edition, in 'hindsight ... the book can be seen as a founding text in the emerging tradition of poststructuralist economic geography' (1998: 942), *Spatial Divisions of Labour* might also be seen as a germinal contribution to (a certain kind of) labour geography. Developing this argument, the chapter revisits both the text and the moment of its consequential intervention, exploring what has passed into practice and what has been lost in translation, before reconsidering the future(s) of labour geography.

Labour, Divided

The publication of *Spatial Divisions of Labour* in 1984 marked an inflection point, rather than simply a catalytic moment, in the double sense that the relevant (intellectual and political) preconditions had been more than a decade in the making, while the full (intellectual and political) repercussions would take just as long to unwind. Let us begin with some of the precursors. Shaking off the legacies of regional science and descriptive industrial geography had earlier yielded a far-reaching critique of location theory (Massey, 1973), followed by a long-term programme of research, mainly in collaboration with Richard Meegan, on the diverse causes and consequences of restructuring in British manufacturing industry (Massey and Meegan, 1982). Shaping what would subsequently be established as the de facto method of economic geography (Barnes *et al.*, 2007), this work both moved with, and responded to, transformative changes in the 'real' economy (aka 'restructuring'). It began – fatefully, as it would turn out – with the multisectoral downturn in UK manufacturing employment after 1968. The language was of Ford rather than Fordism, but it was nevertheless soon evident that this was not merely another case of cyclical adjustment. Neither, however, were the waves of job losses an amorphous outcome of 'industrial decline'. Along with a new generation of industrial geographers (see Sayer, 1982a, 1982b), Massey attacked those analyses that inferred causality from empirically similar outcomes as 'chaotic conceptions', making the case, instead, for a closely argued causal deconstruction of the underlying processes and mediating mechanisms of (employment) change.

Aided by concrete studies of the dynamics of investment, productivity and output changes across 31 industrial sectors, Massey and Meegan concluded that 'job loss' was in fact being driven by three, conceptually distinct processes: productivity-enhancing *intensification* of existing production systems; investment-driven technical change, entailing a reorganisation of these systems; and the *rationalisation* of employment and capacity. This was not hair splitting or analytical sophistry. Its purpose was to isolate the driving causes, socio-economic consequences and intervention points in what was represented as a (qualitatively differentiated) *restructuring* of the production system. The analysis was finely grained, but forcefully articulated nonetheless. It represented, in effect, a different way of writing and thinking what was generally known at the time as industrial geography. And it was politically forthright too. 'We are concerned in this book,' Massey and Meegan explained in *The Anatomy of Job Loss*, 'with the problem of job loss experienced by labour' (1982: 17).

This, needless to say, was a normative as well as a substantive orientation. It was one that influenced Massey's work with the Greater London Enterprise Board, which was at the vanguard of the municipal socialist project around 'local economic strategies' (see Cochrane, 1986; Massey, 1997c). And it shaped the final drafts of what had been provisionally titled *Space and Class*, but which would become *Spatial Divisions of Labour*. What was being lost at the time were not just jobs, but entire regional industrial systems – and along with them, apparently, traditional forms of labour politics and militant capacity. 'This is not a simple rerun of the 1930s,' Massey (1984: 292) said of the impact of more than a decade of intensive restructuring in her own regional birthplace, the North-West of England; 'this time the concatenation of events is taking with it not one but two or three generations of union strength.' The book's publication coincided with the historic tipping point in this process, the miners' strike of 1984–1985, with which Massey was also closely engaged (see Massey and Wainwright, 1985; Hutton, 2005; Beckett and Hencke, 2009). These evidently seismic circumstances, however, were never represented in terms of the inexorable operations of a post-Fordist or post-industrialising economy. The pressing need was for new forms of *political*-economic geography, as the Thatcher government's strategic assaults on the union movement were soon demonstrating for all to see. 'The periods of modernisation [under Wilson in the 1960s] and monetarism [under Thatcher in the early 1980s] were not just different because the wider economic situation had changed so dramatically [or] because the "requirements of accumulation" were different,' Massey contended; they were also 'dramatically contrasting in the dominant political interpretations of what those requirements were' (1984: 266).

Modernisation and monetarism, Massey (1995b: 311) later reflected, were each associated with distinctive forms of uneven development; furthermore, they both reorganised the terrain and the stakes of politics, though they did not (pre)determine the outcomes of political struggles, and nor was it assumed that capital would simply 'get its own way, willy nilly'. The practical and imaginative scope for (labour's) agency was therefore ever-present in the granulated analysis of capital *and class* presented in the book; it may have been calibrated in the context of capital's evolving and variegated strategies, but this never implied capitulation to those strategies. And the analysis of class was always, as it must be, relational. The principal analytical axis was the capital–labour relation, but extending out to regionalised class formations in a way that called particular attention to the diverse intersections of class and gender relations, in particular. This would animate subsequent rounds of locality studies, specifically, but it also paved the way for the feminist economic geographies of the 1990s (McDowell and Massey, 1984; Bowlby, Foord and McDowell, 1986; Massey, 1994a; England and Lawson, 2005). The politics of regional reconstruction were likewise conceived in multidimensional terms, in many ways beyond the class/region nexus.

The evolution of different kinds of spatial structure, their establishment, maintenance and eventual collapse and change, are not simply determined by the characteristics of the labour process, Massey contends, or the requirements of accumulation, the stages of the mode of production, or even the demands of capital. None of these things in themselves 'result in' specific spatial forms. '*Spatial forms are established, reinforced, combated and changed through political and economic strategies and battles on the part of managers, workers and political representatives* ... [They are an] object of political struggle' (Massey, 1984: 85, emphasis added).

Outcomes on the ground, in different parts of the country, could not simply be 'read off' from some rendering of the abstract dynamics of capital, in the critical-realist phraseology of the time (Sayer, 1982b); they would (have to) be understood in terms of 'specific and detailed analysis of the levels and forms of development in different parts of the economy, and of their more specific causes' (Massey, 1984: 81–82). Changes in the labour process, likewise, were not automatically driven by a 'logic of capitalist development'; the outcomes were instead 'case specific and particular', and the overall dynamic 'very much a product of a continual battle between management and labour' (Massey and Meegan, 1982: 184–185).

Not everyone was convinced. David Harvey (1987: 369), in particular, later railed against what he saw as a mode of analysis 'so anxious to deny structuralist leanings or that the "logic of capitalist development" has any explanatory power in local settings ... that all theorising disappears between a mass of contingent labour-management relations in place'. Hinting, albeit obliquely, that there might be a politics of accommodation at work here,[2] Harvey latched on to the 'superficially attractive method' of critical realism as the underlying culprit in what he portrayed as excessively tendentious theorising:

> The problem ... is that there is nothing within it, apart from the judgment of individual researchers, as to what constitutes a special instance to which special processes inhere or as to what contingencies (out of a potentially infinite number) ought to be taken seriously. There is nothing, in short, to guard against the collapse of scientific understandings into a morass of contingencies exhibiting relations and processes special to each unique event. ... Massey ... challenges the direct relevance or power of general theory (always depicted as something 'external') in relation to specificity. Every sentence in *Spatial Divisions of Labour* is so laden down with a rhetoric of contingency, place, and the specificity of history, that the whole guiding thread of Marxian argument is reduced to a set of echoes and reverberations of inert Marxian categories. (Harvey, 1987: 373)

Massey would later counter that, far from being 'inert', Marxian categories were effectively being (re)energised in a form of analysis that was duly attentive – given the demonstrated diversity of conditions on the ground – to the contingent mediation of (immanent) causal powers and concretely realised outcomes, and to the inescapably indeterminate (yet still constitutive) relationship between class and non-class relations (for example concerning gender), in light of the distinctive causal liabilities of the latter.

These were contingencies, local conjunctures and specificities that *mattered*, in other words, and not only to the immediate flow of events or short-term political calculi, but to how processes of economic transformation themselves worked. This did not merely establish 'constraints' or mediating circumstances, but also shaped geographically variable conditions of possibility for both capital *and* labour. It is the *social* sedimentation of the resulting conditions, over time, that was variously summarised as the 'combination of layers', 'locality effects' and the geological metaphor that Massey herself always resisted (see

Massey, 1978, 1991a, 1993b; cf.Warde, 1985; Cochrane, 1987). An explicitly geological rendering of this layering process, of course, might indeed have suggested that the historical categories of analysis were inert, when the logic of the account explicitly called attention to the interaction between layers and the (re)combinatorial, politically mediated nature of the eventual outcomes. Histories of labour-movement docility or militancy, in this sense, did not predetermine the economic destiny of localities, because these 'inherited' conditions were undergoing constant reconstruction in the context of unfolding rounds of investment and disinvestment.This facilitated a dynamic understanding of political-economic conjunctures, while at the same time enabling causal analyses and social imaginaries that, far from being ensnared or constrained by these conjunctures, sought always to exceed them.

A more apt metaphor, in this sense, was the one later proposed by Derek Gregory, who rendered the spatial divisions of labour model in the form of a card game (see Figure 1), in which successive rounds of investment were neatly represented by the four suits, while (local) functions, like assembly or R&D, are indicated by the number of pips. This formulation underlines not only the multiplicity of (local) roles and the (fluid) hierarchies of social relations and production functions in which these are enveloped, but the fact also that different sets of rules are associated with the various games that have been (and might be) played; meanwhile, every (local) player can only work the odds as best they can, with the hand that they were dealt. Local conditions (the succession of hands) also recursively shape extra-local

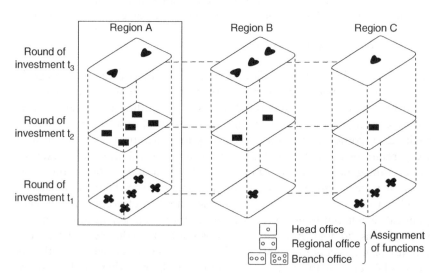

Figure 1   Spatial divisions of labour as card game
Source: Gregory (1989). © Derek Gregory. Used by kind permission

structures and relations, which self-evidently they partly constitute, echoing the sense in which the 'combination of layers signifies a form of mutual determination' (Gregory, 1989: 76). On its face, of course, every card is indeed unique, though at the same time it is (actually and potentially) connected to the other cards according to what can be translated as a range of axes, relations and registers. Economic difference, in the sense suggested by the card-game metaphor, is identified not in the service of separation and sequestration, but in plural forms of relational connectivity – a formulation that would find an echo in Massey's (1991b, 1994a) later arguments around the 'global sense of place'. Peering backwards, distinctive Althusserian residues are also evident in these formulations, in which social phenomena 'are related to each other not primarily through replication or reflection (sameness) but through articulation – the transformative intellectual and social process of creating connections and generating in the process unique beings, situations, and possibilities' (Graham, 1998: 942).

Crucially, while this approach yields an acute (political-economic) sense of the constraints imposed by conjunctural and contextual conditions, in its rendering of the socio-conomic landscape and the 'legacies of the past', it is by the same token an invitation to a different kind of politics of possibility, in that new terrains of political potential are always under construction. So, while the impetus for the *Spatial Division of Labour* may have been to reconceptualise the traditional objects of economic-geographical analysis – jobs, firms, branch-plant regions – in terms of their constitutive social relations, it also made new things of these found objects. 'Rather than as scatterings of isolated phenomena,' Massey later reflected, these 'were to be conceptualised as the interconnected phenomena of social relations stretched out to form a space ... [thereby enabling] us to think the politics of place and uneven development in a different way' (1995b: 326). *Spatial Divisions of Labour*, however, was clearly the beginning, rather than the end, of this process, as Massey's work in the subsequent decade would grapple much more explicitly with global political-economic imaginaries and global senses of place. In the 1970s and 1980s, these had only been manifest in the rather stunted registers of 'international competition' and 'external control'; thinking through the politics of globality in more profoundly relational terms would come later (Massey, 1991b, 2007). The covers of the first and second editions of *Spatial Divisions of Labour* capture this shift quite succinctly: the original yellow cover, featuring a stylised map of the United Kingdom, subdivided by regional-economic functions, is replaced by an abstract and distinctly non-territorial image, evocative of multiplicity and mosaic (see Figure 2). While the cover image of the first edition was credited to Peterborough Development Corporation, Hartlepool Borough Council and Cwmbran Development Corporation, the second edition featured an abstract image courtesy of Russian constructivist artist Lyubov Sergeyevna Popova.

Figure 2a   Doreen Massey, *Spatial Divisions of Labour*, 1984, Macmillan.
Concept © John Hunt. Cover © Palgrave Macmillan. Reproduced with kind permission

Figure 2b   Doreen Massey, *Spatial Divisions of Labour*, second edition, 1995, Macmillan.
© Palgrave Macmillan. Reproduced with kind permission. Cover incorporates L.S. Popova's Representation
of a spatial organisation [construction] (Tretiakov Gallery, Moscow). Cover © Palgrave Macmillan.
Reproduced with kind permission

Recovering Agency

There is, on reflection, some irony in the way that labour geography subsequently emerged, as a self-conscious intellectual and normative project, by way of a less than entirely sympathetic critique of those 'capital-centric' economic geographies that preceded it. The founding claim of proper-noun Labour Geography is that the preceding body of work in economic geography was anchored in a form of structural Marxism that defined the movements of capital and the strategies of firms as the principal engines of change, demoting workers to a secondary, defensive or inert presence in the making of capitalist space (Herod, 1997; Lier, 2007; Rutherford, 2009). Defined, in effect, as a form of analytical and normative corrective, the project of labour geography has taken as its focus, not to say its licence, the processes and practices by which workers '*actively produce economic spaces and scales* in particular ways ... as they implement in the landscape their own spatial fixes' (Herod, 2001: 46, emphasis in original). This call has been a conspicuously productive one, engaging a new genera-tion of activist-scholars in (economic) geography, and connecting to interlocutors in industrial relations, sociology, anthropology, working-class studies, history and other fields in quite new ways.[3] In disciplinary terms, the initiative chimed with the (new) economic geography's ethos of border-less engagement, together with its increasingly energetic spirit of paradigm scepticism and contrarian creativity (see Lee and Wills, 1997). In social and political terms, the rise of labour geography fruitfully coincided not only with the ascendancy of 'post-Seattle' global justice activism, but also with the rise of historically distinctive experiments in service-sector and migrant-worker organising, in community and 'social' unionism, in living-wage and social-rights campaigning, and in networking and capacity-building from the neighbourhood to the transnational scales. Labour geography's forma-tive 'moment' was therefore associated with both propitious intellectual conditions and a (different kind of) generative social base.

This is not the place to review the many and various achievements of labour geography, though through a decade or so of concerted effort these have clearly been considerable. A series of edited collections convey the character of the field: as one typically focused on grounded case studies of labour activism, campaigns, strike actions and organisational innovations, concerned especially with 'new' labour movements of various kinds, broadly rather than narrowly defined (see Herod, 1998; Waterman and Wills, 2001; Bergene, Endresen and Knutsen, 2010; McGrath-Champ, Herod and Rainnie, 2010). It has become axiomatic that 'labour geographers see work-ers as active and capable agents [and] that they have had little truck with the analytical excesses of 1970s structuralism' (Castree, 2007: 855). Very much the same might have been said about *Spatial Divisions of Labour* in its time, of course, with the important qualifier that its (explicit) frame of empirical reference was Britain during its long decade of restructuring (1968–1982), when many 'traditional' forms of labour-movement organisation were collapsing, along with the industrial sectors and regional cultures around which they had historically evolved. This was a period, moreover, in which an increasingly asymmetrical pattern of capital–labour relations had been realigned – first, by the failures of the Labour Party's modernisation project, and then by Thatcherite reassertions of the 'right to manage'. In this sense, the book reflected, if not simply the 'facts on the ground', then certainly a grounded politics, and a grounded political economy. Labour's agency – actual and potential – was certainly not being overlooked, or trivialised, as some side effect of an analytically hard-wired form of structuralist-Marxist myopia. Instead, the analysis sought both to find and make new spaces for progressive labour politics, rooted in but also extending beyond (then) present conditions and capacities.

At the time, left politics in Britain were in the throes of a 'transition' (surely too benign a word, in the circumstances) between an 'old labour' regionalism, which was reaching its historical limits, and an effervescent and confrontational form of 'new left' urbanism, the future of which was still under construction (Massey, 1983a; Massey and Miles, 1984). These opportunities and threats were seen to be co-constituted with a series of economic transformations that were never solely about the steamroller effects of job loss, regional decline, metropolitan restructuring and deindustrialisation, but which generated a 'politics of recomposition' as an object and arena of struggle. In some respects, the landscapes of (tradition-ally) organised labour were literally 'shrinking' (Martin, Sunley and Wills, 1996). In the process, forms of working-class agency were being reconstructed, not least in socio-spatial terms, very much *in relation to* the contemporaneous restructurings of capital and the state.

> [W]e should not simply be seeking the restoration of the old and well-tried. After all, it wasn't a spectacular success. We cannot re-create the old labour movement of the coalfields [which had] its share of disadvantages and vulnerabilities ... There are now different situations, demanding different strategies and forms of organisation. The 'new geography' may look pretty unprepossessing at first sight, but there are possibilities ... The problem is that the movement always seems to be on the receiving end ... never to hold the initiative. The impetus for industrial restructuring has come in an immediate sense from capital. And much of it is a response to, and an attempt to break, established elements of labour movement organisation ... But the fact that that was part of the rationale does not guarantee success. At each end of the process there is now a fight back. (Massey, 1983b: 27)

It was no simple measure of lack of will or imagination that some of these responses – especially in contexts of disinvestment and privatisation – were largely defensive in form; others, however, were both strategic and proactive. Under the banner of 'restructuring for labour', for example, municipal-socialist authorities like Sheffield and the Greater London Council formulated a wide array of 'sectoral' strategies (for 'industries' as diverse as telecommunications, culture, defence and domestic care), less as materially transformative interventions in themselves, more as 'demonstration projects' and 'parables' of socialist practice (Alcock, Cochrane and Lee, 1984; Peck, 2011). These did not seek, naively, to wish into existence capacities for local intervention that did/could not exist, but were developed 'in an attempt to demonstrate a political argument of wider relevance [rather than necessarily] to impact immediately and directly on the local area' (Massey, 1991a: 279; 1997c). These were *political* interventions, inventively positioned within (an analysis of) dynamically restructuring metropolitan economies.

Implicitly or explicitly, this was the methodological-cum-normative logic of the industrial restructuring and locality studies of the 1980s, most of which took *Spatial Divisions of Labour* as a fundamental point of departure (Clark, 1985; Lovering, 1989). The book's substantive content may have been constrained by the landscape of economic crisis in Britain, but its politics – even if they had to begin on this terrain – were not. In one sense, the progressive political potentialities of these moments were always seen as open ended, not least in light of the evident limits of capitalist 'logics' themselves: 'The world is not simply the product of capital's requirements,' the book began; 'there is a lot more determining how we experience space than what "capital" gets up to,' Massey would later reiterate (Massey, 1984: 7; 1991b: 25). Yet at the same time, these spaces of social agency were also defined in a dialectical manner, in relation to the terrains of continuous restructuring themselves, which in some respects prefigured considerations of political strategy but did not ever predetermine them.

If there are premonitions here of what would later become the project of labour geography, there are also some suggestive differences. In its millennial reformulation, labour geography declares simultaneous commitments to labour's 'agency' in the abstract, in normative terms, and in methodological practice. Labour geographers rebuff what they portray as the victimisation script embedded in more orthodox forms of political-economic geography. This has undeniably been a generative manoeuvre, setting in train a project with a militantly contrarian, if not radical, mission. There is no possibility of mistaking whose side labour geography is on: in declaration and in practice, it is has argued the corner of what might be called *re*organised labour. In other respects, however, this orientation has also been self-limiting. Conceptually, the agency-first position has had the unintended consequence of resurrecting a (new) kind of structure–agency binary, in order to locate the project unambiguously on the side of unbounded political possibility. Albeit with the best of progressive intentions, this can have the perverse effect of reifying the very structure–agency binary that so much of restructuring research in the 1980s set out to transcend (Massey, 1985; Urry, 1986; Sayer, 1991), especially if the privileging of agency becomes, in effect, the 'boundary' from which labour geography proper seeks to distinguish itself from its various others and precursors (Castree, 2007; Tufts and Savage, 2009), and if structure and restructuring are somehow defined as analytical no-go areas. The risk is that (all) engagements with 'structure' (however conceived) can end up being tarred with the very broad brush of structural*ism*, when viewed from the other side of the reconstructed wall – one reason, perhaps, for the ill-judged sequestering of the 'spatial divisions of labour' project as another manifestation of capital-centric economic geography.

These are issues, however, that have recently been receiving critical attention from labour geographers, now that the project has apparently entered a phase of consolidation, maturation and reflexive autocritique. Castree (2007: 858), in his sympathetic critique, has for example noted that the signature status of 'agency' has been accompanied by a rather undisciplined form of methodological licence in which the term itself becomes 'a catch-all for *any* instance in which some group of workers undertake *any* sort of action on behalf of themselves or others'. Methodologically, it follows that the socio-intellectual 'productivity' of labour geography – its more-than-the-sum-of-the-parts quality – may be eroded if studies are not conceived in such a way as to yield 'lateral' (cross-case) connections, lessons and insights. What has been identified as a recurring failure to 'set these case studies in context, both theoretically and comparatively', threatens to undermine the explanatory penetration of labour geography, Castree (2007: 858) contends, where the 'failure to distinguish kinds of agency and their enabling/disabling conditions [may lead] to an inability among analysts to say much sensible about worker strategy, normatively speaking'. In practice, labour geography has tended to be 'driven by case studies of successful actions; it tends to focus primarily on unions and union strategies, manufacturing sectors and developed world examples and the notion of 'agency' is oddly rather under-developed and thinly conceptualized' (Coe, Dicken and Hess, 2008: 285).

Steven Tufts (2007: 687) has likewise reflected on the consequences, in the union renewal literature, to trade in 'archetypical narratives', based on something akin to 'best-practice' case studies (see also Lopez, 2004), and the relative paucity of searching, critical analyses of the 'local' embeddedness (versus translocal transferability) of organising success stories or winning campaigns:

> Many studies of labour union renewal also remain largely prescriptive and often 'idealize' labour transformation as an antithesis to the stagnant and defensive actions of retrenched business unionism. As a result, labour union renewal as it 'actually exists' remains hidden as the emphasis is on capacities

to achieve an ideal form, such as a renewed social movement unionism. Labour union renewal, however, contains both complex and contradictory processes with uneven outcomes. In some instances, so-called 'renewed' unionism actually aids the neoliberal project while actions deemed 'business as usual' unionism may forestall its advancement. (Tufts, 2009: 981)

And again, the consequences may be methodological-explanatory as well as normative. Politically, it remains imperative to understand the enabling circumstances, together with the innovative actions, that secured notable labour-movement victories like the Justice for Janitors campaign in Los Angeles, or Baltimore's pioneering achievements in the establishment of living wage ordnances,[4] while *also* probing the vital questions of learning and emulation for this more recent generation of 'parables'. Analytically, there is a self-evident danger that the marriage of an agency-centred ontology and a (radical) best-practices selection strategy can lead to unchecked problems of sampling on the dependent variable. Notwithstanding the political importance of documenting and circulating accounts of strategic or suggestive victories (large or little), in the service of building both capacity and belief, this process is incomplete if such accounts are not effectively complemented with an acute reading of the unfolding political-economic landscape and its associated politics. An appreciation of emergent forms of restructuring – their logics and limits, their conjunctural forms and contingent expressions – need not lead inexorably to the fatalistic suffocation of labour's (potential) agency, or some warmed-over 'new realism'. It might, in fact, serve as a basis for more strategic forms of analysis – grounded in, but at the same time reaching beyond, extant manifestations of mobilisation and politicisation.

These and other concerns have prompted labour geographers to call for a sustained *re-engagement* with the evolving strategies of capital and the state (Castree, 2007; Lier, 2007; Tufts and Savage, 2009; Herod, 2010), indeed with capitalism itself as a variegated and crisis-prone system (Walker, 1999; Wills, 2002; Peck and Theodore, 2007). Curiously enough, this might be seen as something of a return to the conceptual and political problematic of *Spatial Divisions of Labour*, albeit in the context of radically transformed world-historical conditions. It may also represent a belated recognition, of sorts, that this earlier work never represented a capitulation to unbending laws of accumulation, but in its own (situated) way sought to theorise the scope and space for labour's agency in the context of a textured understanding of the grounded realities and emergent potentialities of restructuring. Labour's agency, in this sense, must be understood in *relational* terms, rather than as a one-sided or absolute condition/commitment (Rutherford and Holmes, 2007; Coe and Jordhus-Lier, 2010b). And it must surely be theorised *simultaneously* with nuanced analyses of the restructuring of capital, the state and modes of labour-market governance. These issues cannot simply be 'added back in' as a means of contextualising or correcting a one-sided proclivity for 'sampling on agency'. This must surely entail a much deeper engagement with parallel strands of (political-) economic geography, which have instead been more inclined to 'sample' (and therefore place explanatory emphasis) on moments of intensive restructuring, on conspicuous sites of growth/redevelopment or on episodes of regulatory reinvention. A labour geography that is sequestered from such work represents not just a missed opportunity for mutual enrichment; it runs the risk spawning idealised or even spurious normative-strategic conclusions. To theorise and strategise from the 'restructuring present', however propitious or otherwise these circumstances might be, need not be an exercise in moderation or fatalism. As the political moment, three decades ago, that produced *Spatial Divisions of Labour* showed, these can also be occasions for the development of radical and transcendent visions of labour politics.

The geographies of labour geography must continue to evolve too (see Bergene, Endresen and Knutsen, 2010). Labour's agency no longer simply dwells, if it ever did, in the pores of what used to be known – in a telling but now largely redundant phrase – as the 'advanced industrial capitalism' of the global North. Organisational and political optics derived from these sites (and from their particular labour-movement traditions) may have limited utility, moreover, in a world increasingly shaped by global outsourcing, large-scale migration flows, and radically new spatial divisions of labour. It must be added, however, that *longue durée* analyses of labour's shifting social capacities and patterns of resistance across these global scales hardly present a firm basis for millennial optimism, especially in light of the wars of position waged by key fractions of transnationalising capital.[5] That labour geography should have emerged as a concerted intellectual-political project in the same decade as the effective world labour supply for capitalist production has more than doubled, generating a downward drag on wages and conditions across most of those industries that could legitimately wield the stick of (potential) capital mobility, and pushing unions across the global North into a defensive posture (Freeman, 2006, 2007; IMF, 2007; Berberoglu, 2010), might be read, charitably perhaps, as a reflection of the urgency of the historical moment. But recent pulses of labour activism in China, alongside accelerating labour exploitation and managerial despotism across large swathes of the restructuring landscapes of late-neoliberal capitalism, really underline the need for more meaningfully global (and, for that matter, relational) labour geographies. The field's contrarian posture has certainly been a generative one, but if it is to continue to serve as a guide to both explanation and action, it must also take account of the shifting global landscapes across which labour's praxis is being (re)constructed. Labour geography must also respond to the challenge of not simply finding, but *sustaining*, new sources and modalities of solidarism, in the context of profoundly globalising relations of work, employment and livelihood. Doreen Massey's advice from nearly three decades ago, delivered at an inauspicious moment for trade union politics in Britain, now resounds with distinctly global echoes: 'The labour movement too, if it is to keep ahead of events, must restructure itself, recognise the shifts, address new questions' (1983b: 18).

Turning Points

Intellectually and politically, Doreen Massey was a labour geographer long before the label existed. The path-altering analysis in *Spatial Divisions of Labour* not only marked but also helped *realise* a paradigm shift in economic geography – from positivism and location theory to critical realism and political economy – after which there have been many turns but very few returns. In a quite distinctive way, the not-so-little yellow book that all of us were carrying at the time put labour on the map, not just as an analytical category, but also as a social agent. It did so, however, in a period when long-established forms of labour organisation and politics were being confronted by profound historical challenges – along with certain regionalised fractions of capital and indeed regionalised ways of life with which they had co-evolved. If the book can be said to have marked an intellectual turning point – anticipating the rise of feminist human geographies, instantiating an influential style of relational analysis, and providing some early glimpses of what we would later call globalisation, amongst other things – it was in a more fundamental sense a product of, and a commentary on, a turning point in British economic and political history. In important respects it was born of this conjuncture, while its task was to develop tools for not only understanding but transcending the conjuncture.

The project of labour geography emerged in a quite different age, amid the millennial mobilisations and kaleidoscopic politics of the global justice movement, but also in the context of yet more aggressively globalising and financialising forms of 'contracted out' capitalism (see Wills *et al.*, 2010). With the benefit of hindsight, the moment of deindustrialisation in the global North, two decades earlier, not only coincided with the last gasps of a capital-centric, productivist style of economic geography; it was also the impetus for the first stirrings of a proto-labour geography associated both with a creative reworking of Marxian analytics and with the development, in Britain at least, of politically ambitious visions of 'restructuring for labour'. Massey's contributions, in this context, were diverse and manifold, though certainly one of the legacies was a form of political-economic analysis that called attention to the dialectical nexus of restructuring imperatives *and* strategic openings, of accumulation dynamics *and* progressive political potentialities. Appropriately, in many ways, the project of capital-L labour geography devoted more time in its first decade or so to looking forward and outward, rather than backward and inward. But that now it seems to have

arrived at its own moment of reflexive reassessment, maybe this will extend
to something of a reconstruction of its own antecedents, together with further reflections on the lessons and legacies of earlier conjunctural moments
and turning points. Much has changed since the early 1980s, it hardly need
be said, so some retro-refit would neither be sufficient nor sensible. The
analytical and political challenges associated with *placing and recreating*
labour's agency, in the context of intensified forms of socio-economic and
state restructuring, remain demanding and real ones, however. The
anachronistically styled 'regional problem' having globalised, maybe labour
geography, too, could benefit from looking back in order to grasp its future.

Acknowledgements

I am grateful to Dave Featherstone, Bradon Ellem, Derek Gregory, Hege
Merete Knutsen, Joe Painter, Tod Rutherford, Nik Theodore, Steve Tufts
and Jane Wills for comments and conversation on an earlier version of this
chapter – and to Doreen Massey for inspiration. Responsibility for the
arguments here, of course, is mine.

Notes

1 For surveys and commentary, see Castree *et al.* (2004), Castree (2007), Lier
   (2007), McDowell (2008), Rutherford (2009, 2010), Bergene, Endresen and
   Knutsen (2010), McGrath-Champ, Herod and Rainnie (2010) and Coe and
   Jordhus-Lier (2010a).
2 Somewhat awkwardly, the embrace of critical realism by some left intellectuals
   during the 1980s happened to coincide with a far less principled slide (retreat?)
   into 'new realism' on the part of pragmatic elements in the labour movement
   (see Marsh, 1989; Lovering, 1990; Martin, Sunley and Wills, 1996).
3 For arguments and illustrations, see Castree (2003), Herod, Peck and Wills
   (2003), Mitchell (2005), Ward (2007) and McGrath-Champ, Herod and Rainnie
   (2010).
4 For explorations of these issues, see Schoenberger (2000), Walsh (2000), Martin
   (2001), Aguiar and Ryan (2009) and Wills (2009a).
5 For discussion, see Hobsbawm (1978), Silver (2003), Moody (2004), Seidman
   (2007) and Burawoy (2010).

Chapter Eight
The Political Challenge of Relational Territory

Elena dell'Agnese

Introduction

'Thinking space relationally … has of course been bound up with a wider set of reconceptualisations.' Doreen Massey opens the 2004 special issue of *Geografiska Annaler* devoted to 'the political challenge of relational space' with this claim. She goes on to argue that 'if space is a product of practices, trajectories, interrelations, if we make space through interactions at all levels … then those spatial identities such as places, regions, nations … must be forged in this relational way too, as internally complex, essentially unboundable in any absolute sense, and inevitably historically changing' (Massey, 2004: 5). This poses a philosophical challenge to some of the founding assumptions of modern geopolitical discourse. Alongside place, region and nation we may try in similar vein to reconceptualise 'territory', a notion that is deeply constitutive of the contemporary world political map.

Although it is used rather vaguely in other fields and disciplines (Paasi, 2003; Delaney, 2005), in both mainstream political geography and international relations the term 'territory' refers to a bounded section of space where *nation* states exercise their sovereignty. Since the emergence of the modern state system, 'territory', 'state', 'boundary' and 'sovereignty' have been linked together as mutually defining notions. Thus boundaries have been defined as 'lines that enclose state territories' (Newman, 2003: 123), sovereignty as an attribute 'typically related to a bounded territory' (Paasi, 2003: 113; see also

Spatial Politics: Essays for Doreen Massey, First Edition.
Edited by David Featherstone and Joe Painter.
© 2013 John Wiley & Sons, Ltd. Published 2013 by John Wiley & Sons, Ltd.

Paasi, 2009) and states as 'bordered territorial organizations' (Jackson, 2007: ix). From this standpoint, territory holds 'an epistemological centrality, in that it is understood as absolutely fundamental to modernity' (Agnew, 2009: 746). Indeed, 'across the whole of our modern world, territory [has been] directly linked to sovereignty to mould politics into a fundamentally state-centric social process ...' (Taylor, 1994: 151). Furthermore, while many political analysts have been predicting for almost a decade the death of boundaries and states and the emergence of a flat, geographically homogeneous, postmodern world, the 'modern' concept of territory has a continuing 'allure' (Murphy, 2012) in both international relations praxis and formal geopolitical discourses and theorisations.

Thanks to its 'stickiness', both as a functional device and as a perceptual framework, territory has maintained its role as the most relevant organising principle of today's international system, cornering international relations theory and praxis in a sort of 'territorial trap'. According to Agnew and Corbridge, this trap arises from three geographical assumptions. 'First, state territories have been reified as a set of fixed units of sovereign space. This has served to dehistoricize and decontextualize processes of state formation and disintegration. ... Second, the use of domestic/foreign and national/international polarities has served to obscure the interaction between processes operating at different scales. ... Third, the territorial state has been viewed as existing prior to and as a container of society' (1995: 83–84). Indeed, the 'modern' concept of state territory not only is 'a strategy to control people and to contain society' (Antonsich, 2009: 796), but also a very successful 'semiotic arrangement of space' (Raffestin, 1986, my translation), which evolved during modernity in tandem with the Westphalian state system.

This 'particular ontological determination of space' (Elden, 2005: 16), which has been described as 'abstract, homogeneous and universal in qualities' (Harvey, 1989: 255), is quite the opposite of the one suggested by Doreen Massey when she speaks about relational space. This chapter adopts the particularly stimulating challenge of rethinking territory as a bounded portion of relational space. The chapter traces the genealogy of spatial representations of territory. It will then present alternative uses of the notion, drawn from different perspectives and contexts, to see whether the idea of territory and its boundaries can be reconceptualised from a relational point of view.

Conceptualising Territory

What does a 'conventional' sense of territory really mean? The key discourses through which the political word is made intelligible, such as mainstream international relations, international law, political geography and geopolitics, tend to use the notion unproblematically. So, even if its 'significance … in the modern world cannot be underestimated' (Delaney, 2005: 4), territory has remained 'under-theorised to a remarkable degree' (Elden, 2005: 10) for a very long time.

Introducing his essay *The Significance of Territory* in 1973, Jean Gottmann could write that 'amazingly little' had been published until then 'about the concept of territory, although much speech, ink, and blood have been spilled over territorial disputes' (quoted in Paasi, 2003: 110 and Elden, 2005: 10). However, Gottmann's effort, even if seminal in the field, was not 'as influential as it deserved' (Johnston, 2001: 683), because it did not initiate a tradition of studies. About 30 years later both Paasi and Elden could persist in complaining about the same problem, remarking that 'only a few major studies have been written on territory' (Paasi, 2003: 110) and that 'theorists have largely neglected to define the term, taking it as obvious and not worthy of further investigation' (Elden, 2005: 10; see also Antonsich, 2009: 794). Indeed, apart from the efforts of a few scholars (Gottmann, 1973; Agnew and Corbridge, 1995; Murphy, 2002 and 2012; Elden, 2005, 2009, 2010; Antonsich, 2009), 'territory' is still an object of little reflection both in the field of political geography and in human geography at large. Although Claude Raffestin (1980: 17) describes 'territory' as one of the three distinguishing codes of political geography as a disciplinary discourse (the other two being 'population' and 'sovereignty'), Holloway, Rice and Valentine (2003) do not include it as a 'key concept in geography'. Notions such as 'space', 'time', 'place' and 'scale' deserve two chapters each, and 'landscape' three, but 'territory' does not even make it into the index.

There is no agreement about the etymology of the word 'territory'. The celebrated *Vocabolario Etimologico della Lingua Italiana* (Pianigiani, 1907) says that the term comes from the Latin *territorium*, resulting from the combination of the word *terra* (land) with the termination *-tor, torem*, standing for action – in this sense, *territor* could be accepted as 'the one who owns the land'. Gottmann also puts 'territory' in association with the Latin word *terra*, adding that 'the word territory conveys the notion of an area around a place; it connotes an organisation with an element of centrality …' (1973: 5). Other explanations (Painter, 2006) connect the word with the 'right of terrifying' (*terrere*) possessed by Latin magistrates (in this sense, *territorium* is a place from which people are warned off).

In any case, 'attempts to explain the meaning of the notion through its [Latin] roots are unproductive' (Forsberg, 2003: 12), given that contemporary conceptions of territory developed only later, in modern times. In the medieval statutes of Florence, for instance, the word was apparently referred to as the land belonging to and administered by a town. However, since then 'the concept of territory has steadily evolved' (Gottmann, 1973) together with the movement towards statehood and national sovereignty: 'the history of the modern system of states is a history of defining political power in exclusively territorial terms. The Peace of Westphalia [1648] symbolically marks and normatively enshrines the principle of a sovereignty defined in territorial terms' (Albert, 2001). Westphalia was just the beginning of a lengthy process, since a clear-cut separation of political entities was restricted to the European system of states for a long time. Following decolonisation after the Second World War, the merger of ideas of territoriality and collective identity in the modern *nation* state, with the definition of the principle of national self-determination as a *right* (Albert, 2001), imposed 'territory' as the ordering principle of the political system at the global scale (Badie, 1995).

Territory as a Portion of Space

Even if the notion suffers from 'an absence of theory' (Elden, 2005), in mainstream political geography and international relations the term 'territory' is commonsensically understood as 'a bounded portion of space'. The question is, of course, which kind of space do we have in mind when we make use of the word? Elden contends that territory can be understood 'as a "bounded space" only if "boundaries" and "space" are taken as terms worthy of investigation in their own right as a preliminary step' (Elden, 2010: 13). From this perspective, we can define at least three different notions of space: physical space, geometrical space and relational space, even if they are sometimes confused and partly superimposed. The word 'territory' can be related to all the three meanings and be considered as a portion of physical space, as a portion of measurable/abstract space, and also as a portion of relational space.

Notwithstanding the persistence of geographical myths, such as the age-old invention of 'natural boundaries' (Pounds, 1954; Fall, 2010), the enduring myth of 'landed relationships' (associated with the idea that 'water separates, land connects'; see Connor, 1994), or the assumption that territorial contiguity and cultural homogeneity go hand in hand, physical space in itself has little to do with political geography or international relations. Physical space may be resourceful (since it supplies available land to settle, access to water, minerals and combustible oils, and food); it may enable, or prevent, social and political interaction, providing rivers, plains and seas for transportation, and mountains and other barriers for defence; it may provide proximity between different peoples and states; it may even gain a significance as a symbol of national identity. But its tangible, or intangible, values only exist through the uses that people make of them. In conflicts, it is territory as a physical space that is at stake, and that is fought over, conquered, bombed or walled: again, these actions are not a consequence of its material essence, but only of the multifaceted meanings it acquires in the interaction with human communities.

Territory as a portion of abstract space is more complex to conceptualise, even if it may be defined as the cornerstone of the Westphalian political system. In this case space is assumed to be an isotropic, geometrically measurable and physically divisible flat surface, while territory is usually connected to the ideas of nation – a community of people sharing iconographies, cultures, values – and state – the political organisation expressing the right of self-determination and the sovereignty of the people imagined as a national community. In this way, a portion of terrestrial surface that happens to be fixed as the sovereign territory of a *nation* state is perceived as the nation's homeland, that is as a 'natural division of the earth's surface' (Penrose, 2002: 281) where the members of the nation place their common origins, or at least (as in the case of settler colonies) their common historical experience. As the home of the nation, national territory is generally supposed to be a coherent, homogeneous and ethnically pure 'cultural container' (Taylor, 1994: 155).

The political imagination that views the world as a mosaic of discrete and formally sovereign portions of space (aka 'territories') is usually seen as a product of modernity (Anderson, 1983). However, reconstructing the genealogy of these spatial assumptions and their territorial consequences is far from simple. As Elden remarks in his research on the 'conceptual birth' of the Westphalian system, 'conceptions of geometry and conceptions of territory bear close examination and relation … Essentially, the argument here is that the emergence of a notion of space rests upon a shift in mathematical and philosophical understanding.' 'This development,' he goes on, 'is partnered by a change in conceptions of the state and its territory' and it is connected with the modern notion of measure, as developed by Descartes and by his idea of *Res Extensa* (Elden, 2005: 11–15).

The development of European cartography played a role in the making of this notion of space (Sparke, 2005). In truth, 'cartography apprehends space as pure quantity, abstracted from the qualities of meaning and experience … Such abstraction, objectification, and differentiation are characteristically modern. Cartographic space is analogous to the modern apprehension of time, a quantity [that can be] measured …' (Biggs, 1999: 386–387).

If so, as Farinelli (1994) points out, the history of this notion of space is even lengthier and can be traced, together with the invention of cartography, back to pre-Socratic philosophers such as Anaximander and to Roman cartographers, like Ptolemy. Their legacy disappeared from the Western imagination with the Roman Empire, only to return at the beginning of the fifteenth century with the Italian renaissance, the revaluation and revival of Antiquity, and the rediscovery of Ptolemy's geography. Later on, the development of state geography (Farinelli, 1985) and 'the interplay of cartographic and statistical surveys … established "society" as a field of action, and population [as] defined by the state territory' (Häkli, 2001: 414). Moreover, the explosion of scientific inquiry through the Enlightenment added to the view that 'the world could be controlled and rationally ordered if we could only picture and represent it rightly'. As pointed out by Alliès (1980) and noted by Antonsich (2009: 796), a contribution to this conceptualisation of space was also offered by the jurists of Europe's absolutist monarchies who needed to 'abolish the heterogeneity of places and make them (and the different people who lived within them) equal under the law'.

The process had many spatial consequences. The first is the crystallisation of reality, which is multidimensional and naturally changing, into something flat and static. The second is its simplification: on a map something is there or not there, included or excluded, whereas language allows much more detail. Moreover, maps enjoy the power to divide political spaces, fixing local, national and international boundaries. They function as logo-maps, defining national territories in easily recognisable shapes (Anderson, 1991). They inscribe names, shapes and signs, turning those portions of space into symbolic landscapes (Sturani, 2008). Together with statistics, they connect space to state government. Thus, the geo-metrification of the world through its cartographical representation sacrificed its fluidity and variety in exchange for the possibility of dividing and dominating it.

From this standpoint, territory acquires a strong symbolic importance (Murphy, 1990), even if regimes of territorial legitimation tend to confuse its historical attributes with physical qualities. Indeed, in practical geopolitics the two ideas of territory as a fraction of material reality and a portion of abstract space are strictly interrelated (Newman, 1999). But they both acquire meaning only if inserted in, or as a result of, a sum of interactions between individuals, contexts and societies.

Territory as a Portion of Relational Space

Practical geopolitics rarely questions the quintessence of the world system as composed by a given number of separate sovereign territories, succumbing to 'the inertia of modernist territorial ideas and assumptions' (Murphy, 2012). The theoretical assumptions of this regime of truth in the last two decades, however, have been variously disputed in different disciplinary contexts and from different points of view. In geography Amin (2004: 33) has remarked that 'the first challenge is the rise of compositional forces which are transforming cities and regions into sites immersed in global networks ...'. Massey's work has been central to this reimagining of spatial vocabularies. Her work on cities, for example, has drawn attention to their character as 'open intensities' produced through the distinctive constellation of relations that are brought together in different urban sites (Massey, 1999d: 161). This opens up important challenges for the production of territory through thinking about how such changing connections beyond cities are inimically related to contestation and power relations within cities.

More generally, counterposing networks (of flows, goods, capitals, ideas) and territories has been highlighted by many social theorists (see Painter, 2006), variously suggesting a vision of a more and more interconnected planet where globalising forces were fusing together, and somehow cancelling or intermingling cultural specificities and national characters, as well as territories (Badie, 1995). Even if some of these observations have later been subjected to a thoughtful reconsideration (Newman, 2006), the dichotomous opposition between a 'modern' world of fixed, and static, territories, and a dynamic, unstable and 'postmodern' world of flows and networks has become a sort of refrain.

A closer observation, however, may suggest that even this position is not totally satisfying. A first critique might focus on the temporal dimension through which the dichotomy is interpreted. Indeed, seen from the perspective of Jean Gottmann's classic model of iconography and circulation (Gottmann, 1951, 1952) we cannot say that territories were once properly bordered cultural containers, while nowadays they tend to be swamped by international networks and global fluxes. On the contrary, the political partitioning of our political world has always been shaped and reshaped by the opposition of the contrasting forces of iconography (cultural traditions), which tend to root people inside stable political communities, and circulation (fluxes and networks), which unsettle traditional communities, creating opportunities for interaction and change.

Joe Painter develops a more substantial critique, arguing that the opposition between territory and network is nothing new – and furthermore, that it does not even exist: there is not a 'territorial entity' which can be opposed to a system of networks, since 'territory' is a network in disguise:

> territory as such has no real existence ... Territory is rather an effect of networks ... The configurations of practices and objects, energy and matter that go by the name 'territory' are no more and no less than another set of networks ... Territory is not a kind of independent variable in social and political life. Rather, it is itself dependent on the rhizomatic connections that constitute all putatively territorial organisations, institutions and actors. (Painter, 2006: 28)

Such a statement is from one side very provocative, from the other apparently dangerous. It is provocative because it dispels the old notion of 'territory' as a portion of measurable and abstract space. It is dangerous because, if we accept the idea that it is just a bundle of networks, the very concept of 'territory' seems set to lose its own right to exist as a specific concept.

In order not to throw the baby out with the bath water, a useful hint on how to rethink 'territory' as a cluster of interrelations may be found in the work of the French geographer Claude Raffestin, who attempts to situate explanations of 'territory' and 'territoriality' within the reconceptualisation of power provided by Foucault. Specifically, Raffestin (1978) defines territoriality as a network of relations, which originate from a tri-dimensional system based on society–space–time. In this perspective each individual develops their own relation with the context, in a formula that can be summarised as HrE (where H is the individual, r his specific relation and E the context). 'Territoriality' can be expressed as $T\Sigma HrE$: that is, as the sum of each individual's relation with the context. 'Context' is composed of material as well as social (or relational) space, while 'territory' can be considered as 'a way of giving order to space, which must be understood inside the system of knowledge of a human being, as a cultural subject' (Raffestin, 1980, my translation).

As Klauser (n.d.) remarks, 'this approach strongly differs with the widely held understanding of territory as the sphere of influence of political authorities, mainly in the form of the nation state. Raffestin, on the contrary, argues that territory must be seen more generally as socially appropriated space, which is closely interconnected with society on different spatial, temporal and social scales.' All the same, the apparent difference can be bridged if we accept the general assumption that the state represents just one of the many possible forms of human political association.

Indeed, it may become even less relevant if we combine Raffestin's ideas, and specifically the concept of territoriality as the sum of interactions that each of us develops with context, with Doreen Massey's relational understanding of space. Thanks to Massey's ideas, this reconceptualisation of territory 'as part of a reciprocal relationship with society … as a processual and relational, socially produced reality' (Klauser, n.d.), may acquire a new political meaning, which may help us override the 'violent cartographies' (Shapiro, 1997) produced by the old notion of territory as a cultural container.

If territory is not a cultural container, homogeneous on the inside and separated from the outside, but a portion of relational space, porous, processual and unstable, the usual exercise of mapping geopolitical dangers, drawing boundaries between 'us', the friends, and 'them', the Others, ontologically different and potentially enemies, turns out to be extraordinarily pointless. If differences are not to be essentialised and spaces are not fixed but relational, if we accept the idea that spatial identities too are relational and not fixed, we can also accept the idea that being a friend, or an enemy, does not depend on where you were born or where you live. Rather, it depends on the practices of interactions ('which include non-relations, absences and hiatuses' [Massey, 2004: 5]) you succeed in forging together – that is, in the sum of individual relations with the context.

Territory as a Bounded Portion of Relational Space

If territories are portions of relational space, and not portions of abstract homogenised spaces, the quality of their interactions is not an (inescapable) outcome of the essentialised characteristics of homogenised populations, but a consequence of the sum of interactions within, and among, individuals. In international relations, this assumption brings some practical effects. For instance, during the war in ex-Yugoslavia in the 1990s it could have helped us to understand that the reason for the conflict was not to be found in the violent characters of the local populations, or in supposedly 'ancestral hatreds', as many media commentators and political analysts argued at the time (dell'Agnese and Squarcina, 2002), but in the antagonistic interests of groups and individuals. This emphasises how identities which were interpreted and temporalised as 'ancestral hatreds' need to be understood as part of contemporaneous, ongoing relations (Massey, 2005: 119–120).

A relational understanding of space would have suggested that the solution to the conflict lay not in the creation of partitioned territories but in the weaving of more positive interactions. Today, a similar understanding could make us aware that, if someone decides to become a terrorist, it is not simply because they were born inside the wrong territory, but because their interactions with the context – no matter the scale – went wrong. This means, firstly, that we will not defeat terrorism by waging war against it, that is, by bombing other people's territories. In fact, by following that strategy we will make things worse, since the qualities of our relations and interconnections with the inhabitants of those territories will deteriorate. Secondly, it suggests that we cannot defend ourselves by building walls, since terrorism, being a relational problem and not a territorial one, can develop inside any kind of relational context. Thirdly, it implies that, since we are not living in separated societies but in interconnected systems of relations, or, better, in a global/local system of power-geometries, as Doreen Massey (1993a) would say, we are always part of the context, and for this reason we have to take our own responsibilities seriously. This holds even though it is plainly easier to recognise our involvement at the local scale and much more difficult to do so when we are apparently far away from the scene of the action.

Thinking space relationally, and believing that we are interconnected at the global level, does not mean that scale is irrelevant or boundaries are immaterial. In this context we can follow Doreen Massey, when she affirms that we are not faced with 'non-striated smooth space', an undifferentiated global network for a global multitude (Massey et al., 2009: 417). On the contrary, 'striations do exist', places still exist, and recognising them may be helpful. 'What we have to do is take responsibility for the striations that do exist. Take responsibility for the boundaries, the definitions, the categorisations.' And she goes on: 'Although conceptually I believe definitions are necessary, and that sometimes we have to draw boundaries in space, the job is to take responsibility for the drawing of those lines on the one hand, but also to recognise that the social relations which cross those boundaries aren't only ones of antagonism.'

So, we are again faced with bounded portions of space, which can be defined as territories. From this perspective, the old notion of territory needs to be reconceptualised, but may turn out to be useful again. Indeed, the modernist notion of territory can be accepted as a 'way of giving order to space' (Raffestin, 1986): to this perceptual and functional rationale it owes its long-standing fortune, but also its contemporary stickiness. We cannot erase the connection between 'territory' and 'boundaries'. But we must reconceptualise both of them, rethinking territory as a bounded portion of relational space, and boundaries as a tool to organise those relations, not simply a 'language and an imagination of othering'. We have responsibility for the quality of the borders we draw because if conflicts and terrorism or peace and collaboration are going to be established between different territories, it is not a question of walls. To conclude, we can go back to Doreen Massey (Massey et al., 2009: 417), who states: 'So I have now got myself into a position where I am wanting to say a global sense of place absolutely yes, but that doesn't mean "abandon place".' Nor abandon territory, we might add: learning how to think relationally about it may indeed prove to be a powerful political challenge.

Interlude
Your Gravitational Now
Olafur Eliasson

I make my day by sensing it. Measuring by moving, my body is my brain. My senses are my experiential guides – they generate my innermost awareness of time while generously giving depth to my surroundings. Constantly and critically invested in the world of today, they receive, evaluate and produce my reality.

When I walk or drive through the Icelandic landscape, I sense the surroundings and sense myself searching for sense. This vast landscape is like a test site that nurtures ideas and helps me process them into *felt* feelings – maybe even into art. Exercising physical and perceptual means of charting out space, of *becoming*, is for me a way of speaking to the world. This method or 'technique' raises questions that might just as easily be asked at different times in different situations, removed from their art context. Depth, time, psychological and physical engagement, perception – topics abound for which the landscape welcomingly offers experimental conditions and material.

In Iceland and elsewhere, I continuously exchange my private being for a shared reality. I – sensorium, feelings, memories, convictions, values, thoughts, uncertainties – only am in relation to the collective.

Imagine standing on the vast banks of black sand just south of Vatnajökull, the largest glacier in Iceland, looking northward onto the tip of Skeiðarárjökull, one of its glacier tongues (Figures 3 and 4). From this particular point, the wide glacier takes up a large part of the horizon, and its gravel- and ash-covered nose sprawls into an ungraspable mass. Abstraction

Spatial Politics: Essays for Doreen Massey, First Edition.
Edited by David Featherstone and Joe Painter.
© 2013 John Wiley & Sons, Ltd. Published 2013 by John Wiley & Sons, Ltd.

Figure 3 Olafur Eliasson, *The glacier series*, 1999. 42 C-prints. Series: 244 × 404 cm, each: 34 × 50 cm. Photographer: Jens Ziehe. The Solomon R. Guggenheim Museum, New York. © 1999 Olafur Eliasson

Figure 4   Olafur Eliasson, *The glacier series*, 1999 (detail). 42 C-prints. Series: 244 × 404 cm, each: 34 × 50 cm. Photographer: Jens Ziehe. The Solomon R. Guggenheim Museum, New York.
© 1999 Olafur Eliasson

and impalpability pervade, filtered through your here-and-now body. Standing right in front of the glacier, you may first begin to feel a degree of intimacy and familiarity. The experience of proceeding onto the glacier itself is a moment of intense physical drama. Pressurised by the mass of ice, a sub-glacial water current causes the otherwise dry black sand right in front of the tongue to undulate like a fatigued trampoline. Cautiously trying to cross the few yards of billowing sandy surface to the glacier itself, you develop a funny, anti-gravity-like gait – a bit like moon-walking. Hoping to defy physics, you make yourself light, distribute your weight as evenly as possible, heart pounding. Quicksand below threatens to pull you in.

Three years ago, together with a local driver and my good friend, the landscape architect Günther Vogt, I undertook a trip on Skeiðarárjökull that began with this chillingly destabilising experience. Our journey was charted by glacier mills, those incredibly deep holes in the ice carved by the rush of surface glacial melt water and debris (Figure 5). For two days we travelled from void to void. Strapped to a ladder cantilevering off the roof of an all-terrain vehicle, I would lie suspended horizontally in the air, camera in hand, examining this extraordinary phenomenon directly from above, these ice perforations, each unique in shape, depth and balance of dirty grey, white and turquoise hues. Add to this a powerful soundtrack of the rushing melt

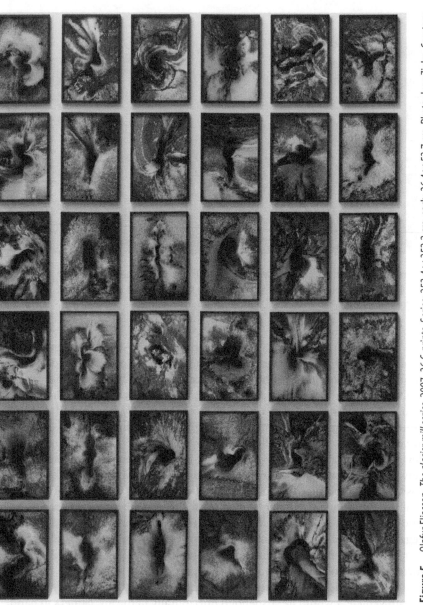

Figure 5 Olafur Eliasson, *The glacier mill series*, 2007. 36 C-prints. Series: 253.4 × 352.2 cm, each: 36.4 × 53.7 cm. Photo: Jens Ziehe. Courtesy the artist; Tanya Bonakdar Gallery, New York; and neugerriemschneider, Berlin. © 2007 Olafur Eliasson

Figure 6   Olafur Eliasson, *Iceland series*, 2007. C-print, unique. 60 × 90 cm.
© 2007 Olafur Eliasson

water spilling into the voids. The mills made explicit this wonderful interior life of the glacier, its transformations, inner crackling and grinding, and other sub-surface noises that we tend to associate with the sounds of deep oceanic life. They were like glacial loudspeakers, whose sound waves spoke of the type of dark space I was studying from above. My – sadly soundless – archive of photographs later became *The glacier mill series* (2007).

Once on my feet again, walking on the crusty crystalline ice surrounding the mills, I would gaze southwards over Skeiðarárjökull and the sandy lowland, towards the coastline and horizon in the distance. This is a perfect set-up for an exercise of the senses. On the large surface of the glacier, no familiar forms or objects give a sense of scale or distance. The ice is like a perfect, giant, tilted plane as far as the eyes can reach. If you halt, the sheer size of the ice plane blots out its subtle sloping towards sea level, about 200 to 300 metres below. Looking firmly towards the horizon, you suddenly experience the large icy surface as if it were raised to become horizontal with the sea beyond, tipping surrealistically towards you (Figure 6). For a second, the rational structuring of your perceptions is short-circuited. All knowledge of space and its dimensions dissolves. Breathtaking! The brain protests: obviously it is the glacier that is inclined, not the black sand desert and the sea. But the vivid unfamiliarity of the situation makes it surprisingly difficult to ascertain which of the two is inclined. Should you consciously decide to see it that way, the image can snap back into a sloping glacier with flat land and horizon. The exercise consists of this sense/brain-driven oscillation, a peculiar back-and-forth between self and surroundings.

When I started walking again, the ice gloriously crackling under my feet, my vestibular system at work, I found myself adjusting my steps to even minuscule changes of level in this sprawling glacier landscape. Gravity, movement and the passing of time effortlessly conjured up the feeling of the inclination of the ground. Then the slightly oblique surface of the glacier became immediately graspable again. Consider this simple sensory re-evaluation, prompted by walking and driving around on the great Vatnajökull, a mild attack on the ever-increasing experiential numbness that society produces in abundance. Evaluating the (partially self-imposed) rules by which we live implies giving renewed attention to our definitions of time and space, and of ourselves as sentient agents.

At my studio in Berlin I work with similar sensory-motor experiments. The sandy underground of this city, vaguely reminiscent of the Icelandic quicksand zone, continues to challenge local city planners and inhabitants. 'Berl', derived from Old Slavic, probably means 'swamp' – the swamp upon which Berlin was built and which has influenced its urban organisation. Often less than three metres below the surface, the natural groundwater level calls for unusual building methods. When Potsdamer Platz was rebuilt following the fall of the Berlin wall, deep giant holes were made in preparation for the high-rises to come; and, had the construction sites not been sufficiently sealed off, the adjacent Tiergarten park would have been threatened by dramatic changes in its hydro-geological balance, causing trees and shrubs to die.

My studio sits on a hill called Prenzlauer Berg. Here the swamp has given way to a stable foundation. Originally a brewery, the building is equipped with a maze of double-tiered basements for beer storage. In 2009 I opened a school, called the Institut für Raumexperimente, and the first exhibition by the students took place in May 2010, in the cold but dry basements below – its title: 'Let's start to implement little errors'.

Imagine leaving the studio with me and turning left onto Christinenstrasse, which slopes down to the intersecting Torstrasse about four blocks south and ten metres below your current position. Surprisingly, an exercise similar to the one on Skeiðarárjökull can be made in this street, its one side flanked by the playgrounds on Teutoburger Platz, the other by the traditional Wilhelminian buildings of Berlin. The experiment is perfect for an unpretentious afternoon with students. Stand still, focusing on Torstrasse. Try to conjure up a sense of the street sloping softly downhill in front of you, the houses, the sky, the light, the other people. Now use your imagination to lift up the street to horizontal. Simply tell yourself that the slope in front is in fact level. Freezing this image, you notice that the buildings lean curiously towards you. At the end of Christinenstrasse, Torstrasse is now subtly tilted, fitting the image you have produced. Enjoy the thrill of this image or examine your feeling of what physical impact the spaces of the city have on your way of sensing them and you in them. By an act of sheer will, flip the image around, once again letting Christinenstrasse roll naturally downhill, the buildings straightening themselves to upright position. The brain finds it unexpectedly easy to make such momentary, relative contracts with reality. Our senses and surroundings are easily manipulated.

Start walking down the street, registering the inclination with your feet. You may experience a minor fall with every step you take – and another thrill while your limbs jiggle slightly. When in motion, our bodies co-produce what we sense, partially handing over the production of the space-so-far to our feet, eyes and entire sensorium. The second exercise is to experience the difference between simply walking down the hill and walking while imagining Christinenstrasse to be straight. The discrepancy between the physical registering of the street and your brain projection imbues your experience of the street with a *felt presence*. The mental exercise disrupts and reshapes this everyday functioning of our senses, of our selves, our urban surroundings – exhibiting the sensory numbness (indirectly) nurtured by many city planners.

The intricate translation of information among our limbs, brain, perceptual apparatus and sense of orientation is a vehicle of the self. To embark on such exercises and journeys allows time to give space to feelings. Walking becomes a tool for *emotionalising* space, a landscape, an urban setting or a building.

Claude Parent and Paul Virilio, the French visionary architects of the 1960s and founders of Architecture Principe, celebrated *the function of the oblique* – moving and living on inclined surfaces, the body almost always tense, in constant disequilibrium or only momentary balance – as a vital spatial principle that could heighten the quality of life. What this also does is it counters the hegemonic status of vision (as if separate from the other senses) in a society obsessed with images, their mediation, and the representation of power. It is based on a belief in time, in transformation, and the potential for radical change.

Today, I insist on a similar kind of spatio-sensory holism, where art can challenge and change societies by instantiating different relations to the world, where actions and consequences matter; art takes seriously the space-producing abilities of our bodies. It prompts us to re-evaluate the value systems according to which we measure ourselves and our surroundings; it insists on friction and difference. This offsets the alarming sugar-coating of experiences developed by a world that (involuntarily) generates numbness, a world obsessed with profit and consumerism, which packages experiences for sale rather than insists on individual and collective responsibility for sensation and shared space. As Barbara Maria Stafford has put it: 'Hiding the mechanisms behind visual construction is like window-shopping. No longer seeing the constitutive technology encourages the ingestion of a seamless spectacle of goods' (Stafford, 2007). In this type of spectacle-driven world, our ability for refined sensation is dulled. Navigation happens via GPS, according to a map that represents a world 'out there', rather than with a map that we draw as we go – what Bruno Latour calls a 'dashboard ... or a calculation interface that allows you to pinpoint successive sign posts while you move through the world' (November, Camacho-Hübner and Latour, 2010: 595). Ultimately, I am trying to produce sense.

Epilogue

I met Doreen at a lecture on walking by the artist Hamish Fulton, while I was preparing *The weather project*, which was later installed in the Turbine Hall of Tate Modern. This marked a turning point for me. I had for some time been interested in duration, temporality, and how our experience of time co-produces space – topics that are at the core of my artworks. Where phenomenology, which had been decisive in my early work, addresses temporality from the singular perspective of a subject, Doreen insisted on thinking of the subject contextually. Imagine a person boarding a train in Manchester, going to Liverpool, and disembarking at the station:

> Your arrival in Lime Street, when you step off the train, begin to get into the things you came here to do, is a meeting-up of trajectories as you entangle yourself in stories that began before you arrived. This is not the arrival of an active voyager upon an awaiting passive destination but an intertwining of ongoing trajectories from which something new may emerge. Movement, encounter and the making of relationships take time. (Massey, 2003: 110)

Later we would talk more about the subject in relation to its social surroundings, the performative collectives in which it participates. For me, sensitivity to the mutable social context became a topic I developed while occasionally crossing paths with Doreen. I benefited from her belief in making explicit the changing conditions under which exchanges take place and movements are made, conditions that always co-produce internal and external performances. Doreen has changed my way of seeing my work in the world and the world in my work.

Part Three
Reconceptualising Place

Chapter Nine
Place and Politics

Jane Wills

Introduction

On 3 May 2010, three days before a general election, more than 2000
people assembled at Methodist Central Hall, London, to engage the
leaders of the three main political parties. Whoever was to hold the
balance of power after the election, the organised citizens of the UK were
determined that they would already have a place at the table. Each party
leader was asked to respond to the Citizens UK manifesto to support the
living wage in publicly procured jobs, a cap on interest rates set at no
more than 20 per cent, community land trusts as the basis for the development of affordable housing, an earned regularisation for long-term
irregular migrants and an end to child detention in immigration centres.
Each was also asked to meet Citizens UK at least once a year and to
attend at least two civic assemblies over the course of the next parliament.
In the words of the programme: 'The voice of the people should be heard
by our elected leaders.'

Faced with more than 2000 people, the media and the desire for votes,
each party leader agreed to most, if not all, of the manifesto, as well as
agreeing to meet on an ongoing basis. Since the election, representatives
from Citizens UK have been meeting with government ministers and their
advisers to discuss the living wage (involving the Treasury and the
Department for Work and Pensions) and child detention (with the Home

Spatial Politics: Essays for Doreen Massey, First Edition.
Edited by David Featherstone and Joe Painter.
© 2013 John Wiley & Sons, Ltd. Published 2013 by John Wiley & Sons, Ltd.

Office and Deputy Prime Minister Nick Clegg's team). Prime Minister David Cameron will be held to account for his promise to support community land trusts and to attend at least two civic assemblies during the years before the next parliament.

Citizens UK comprises a broad-based alliance of civil society organisations, including faith, labour, educational and community organisations working in alliances in London (as East, South, West and North London Citizens), in Milton Keynes (as Citizens MK) and as Citizens for Sanctuary in the Midlands, Tyne and Wear, Nottingham, Manchester, Plymouth and Solent. Drawing on the experience of the Industrial Areas Foundation in the USA, Citizens UK uses the experience of training and political campaigns to foster connections between different groups, to develop leadership and to teach politics. By building solidarity on the basis of shared geography, the organisation has been able to arrange meetings with the national political party leaders, end the practice of child detention, secure a living wage for more than 10,000 low-paid Londoners, foster initiatives to foster community safety and inch towards a community land trust in Tower Hamlets.

At a time when economic globalisation, increased population mobility and consumerist individualism are eroding the connections between people and places, the experience of Citizens UK seems oddly out of sync with our age. Some would argue that the organisation represents the last gasp of the old order that will inevitably be replaced by more mediated experiences of politics engaging voters as isolated individuals sitting in front of a screen. Many would suggest that the geographically structured politics of democratic engagement can only decline in a world where people's connections and interests already spread across borders – to 'home' in the case of migrants or 'the planet' in the case of environmental campaigners. My own experience, however, suggests that this view is too hasty a rejection of the importance of place. We all live in places even if we elect to commute further, to move more regularly and/or with a greater degree of choice than we did in the past (Savage, 2008). Almost all of us are dependent upon the state-funded services provided in our local communities and we elect our local, national and regional political representatives on the basis of place. We are also acutely aware of the degree to which we live in civil communities, where people respect and have connections to each other. Our local schools, universities, hospitals, councils, churches and mosques – amongst others – provide important geographical anchors in our everyday lives. The experience of Citizens UK suggests that these institutions provide the spaces from which to re-engage people in politics, however mobile those people might be (Wills, 2012).

At this conjuncture, and despite the time-honoured connection between the polis and the emergence of democracy, geographers seem to have remarkably little to say about these connections between place and politics. Indeed, the hegemonic ideas of the discipline have reinforced the idea that place is less significant than it was in the past. Places are understood to be intersections or nodes in spaces of flows, in which capital, people and ideas are constantly moving. Places are seen as unbound and fluid, providing weak foundations for political practice. Most clearly associated with the work of Ash Amin (2002b, 2006, 2010) and Nigel Thrift (Amin and Thrift, 2002), places are characterised as sites for occasional fleeting encounters where lasting face-to-face relationships are a thing of the past. As Amin (2010: 55) has recently suggested, in a rather bleak prognosis for the future of place-based community:

> Today, people live next to each other largely as strangers, in places that hardly hold together as communities of common fate or interest, without much contact with each other, often moving on to live elsewhere. Individuals have real-time and intimate contact with people and things far away, and so dwell in disparate and often physically distant worlds of affiliation and feeling ... the neighbour is just the person next door and neighbouring is no longer a required art of living.

In this context, local politics is generally viewed with suspicion as a product of sentimental nostalgic identifications, exclusionary communitarianism and/or hostility to outsiders. There is little sense that local political identifications can be nurtured and mobilised for progressive reform.

In this chapter I want to explore the connections between place and politics. In so doing, I focus on Doreen's key contributions to these debates, differentiating two distinct phases in her research. While Doreen herself has never differentiated her work in this way, the time-spaces in which she developed her analysis were different, and her engagement in the political battles of the day is inevitably reflected in her work. In the first, dominant in the 1980s, academics and activists were able to use particular places to launch political campaigns in opposition to the emerging neoliberal order pioneered in the Thatcher years. Doreen was at the forefront of this political agenda – working with the Greater London Council – and I describe her writing from this period as her focus on 'politics *in* place'. The second, developed since the 1990s, was developed during the entrenchment of neoliberal globalisation, when it was politi-cally important to stress alternative forms of international connection. I call this Doreen's focus on 'the politics *of* place'. In what follows, I explore each in turn before going on to suggest that the recent shift towards understanding the politics *of* place has somewhat overshadowed some of the important insights of Doreen's earlier arguments about politics *in* place. While this is an issue about the way in which Doreen's work has been used as much as the words she has written, I argue that there is now much to be gained from a closer integration of the two sets of arguments in future research.

Politics in Place

Geographers have long struggled to articulate the relationship between the local (the particular) and wider social processes and relations (the general). The discipline requires that we grasp the importance of place and its specificity while also being attendant to the wider social, economic and political processes which make and remake places and the geographic relations between them. Historically, the pendulum has swung between the poles of this continuum, moving from a focus on the particular towards more abstract sets of concerns. Thus, a cultural geography of the region – popular in the 1950s – came to be overshadowed by analyses of the laws of neoclassical and/or capitalist economy during the 1960s and 1970s, while attention has focused back towards local specificity in more recent times (Castree, 2009).

Doreen has been at the forefront of the discipline from the 1970s, and part of her brilliance has been to avoid either end of the spectrum. While deploying the new and sophisticated lexicon of Marxism, she sought to connect particular places to the wider social processes of which they were part. First developed in her geo-comparative research into the processes of political-economic restructuring across Britain's coalfields and in Cornwall (Massey, 1983c), and in her work on spatialised gender differences in labour market participation with Linda McDowell (1984), her approach was most clearly articulated in *Spatial Divisions of Labour* (1995 [1984]; see also Peck, this volume). In brief, Doreen argued that places were being differentially implicated in, and transformed by, the processes of economic restructuring underway at that time. As capitalism changed, so, too, particular places would inevitably be repositioned in a reconfigured and necessarily spatial division of labour. Doreen developed a geographical analysis of the world that was changing around her. The dramatic decline of manufacturing industry, the growing power of finance capital and fierce contestation over the speed and direction of economic and political restructuring highlighted the uneven geography of change and the political importance of places as sites for orchestrating and resisting such change. Doreen famously described how capital deposits itself unevenly across the landscape in successive layers of investment, thereby remaking places and the socio-political opportunities of the people who live in them. Massey's idea of the deposition and combination of layers of capital – and capital's subsequent relocation – in the relentless search for profit became one of the chief ways in which geographers came to understand the nature of place, the connections between different places and the spatiality of the political opportunities that exist to engage with and/or resist such change.

Doreen's ideas were widely taken up by scholars across the social sciences during the late 1980s, culminating in the ESRC-funded localities programme, directed by Phil Cooke and comprising seven locally oriented research projects (Cooke, 1989; Massey, 1991a).[1] Taking direct inspiration from Doreen's previous work, the programme was designed to explore the particularities of economic and political change in a number of towns and cities in the UK while also understanding each place to be a manifestation of common shifts in the nature of capitalism and the spatial division of labour. In sum, this body of work demonstrated that 'geography matters' in relation to the direction and impact of political-economic restructuring and, as such, these local differences are then constitutive of the wider processes of change that take place.

In relation to politics, this phase of Doreen's work highlighted the geographical differentiation of popular traditions of resistance. While capital made and remade places, local people were also active, creating and sustaining their own political organisations which could form the bedrock of any resistance and/or the development of alternatives. As she put it, 'people in different parts of the country had distinct traditions and resources to draw on in their interpretation of, and their response to, these changes [in political-economy]' (Massey, 1991a: 268). Citing the importance of geographically rooted political organisation in the development of the labour movement in the UK, Doreen argued that places like Glasgow, East London, Yorkshire and South Wales had resilient political traditions and shared experiences of struggle which could facilitate further resistance to unemployment, inequality and injustice. Indeed, her own experience of working with the Greater London Council demonstrated the importance of popular engagement in locally respected political institutions when mounting resistance and developing alternatives to Thatcherism. As Peck (this volume) suggests, Doreen's hands-on engagement with local politics in London was as much about reorienting and reconfiguring the left as it was about simply rehearsing traditional forms of political practice.

Although it is rarely acknowledged, this work strongly resonates with the subsequent development of what has become known as 'labour geography'. Doreen's arguments about the spatial division of labour and the geography of political organisation played a very significant part in the development of this new stream of research (Massey and Miles, 1984; Massey and Painter, 1989). As a new generation of geographers (a number of Doreen's doctoral students amongst them) sought to make sense of the political possibilities that existed in the wake of the rout of the labour movement that took place in the UK during the 1980s and 1990s, they were drawn to explore new service-sector trade unionism, labour-oriented urban social movements and experiments in labour internationalism that were starting to develop (Herod, 1998, 2001; Waterman and Wills, 2001; Hale and Wills, 2005). Doreen's scholarship provided a foundation for this new wave of research that highlighted the importance of trade union traditions and workers' collective organisation in resisting the power of capital and seeking to make space and to spatialise strategy in the interests of labour (see also Herod, Peck and Wills, 2003; Peck, this volume).

The Politics of Place

In the 25 years since *Spatial Divisions of Labour* was published, neoliberal globalisation has extended and deepened the interconnections between places across the globe. Reflecting the growing importance of this traffic between places, Doreen's more recent work has been part of a wider shift towards relational thinking in geography (see also Amin, 2002a, 2004). Most clearly explicated in *For Space* (2005) and *World City* (2007), Doreen has highlighted the social, political and economic relationships that create place. In doing so, she has explored what I have termed the politics *of* place. By this, I mean the unequal power relations that structure our world, its places and the lives of its people. Thus in relation to London, Doreen has highlighted the role of this city in the development of a global neoliberal political-economy that, in turn, has facilitated the further growth of the city on the back of a myriad of relations of exploitation and uneven power relations that stretch across the globe. The power-geometry of London's relations with the rest of the world have fuelled the economic growth of the city – and, particularly, the wealth of its richest inhabitants – to the detriment of many of the poorest in the city and the rest of the world. Citing Mike Davis's work on the exponential growth of urban slums in the global South, Doreen has argued that London is 'a powerful part of the same dynamics that produce, elsewhere in other cities, Davis' "planet of slums"' (Massey, 2007: 9–10).

This phase of Doreen's work has urged us to think about the politics that lie behind the formation of place, and, in turn, she calls upon us to address our spatial responsibilities. Thus, in *World City*, she advocates taking responsibility for the way in which London's health services depend upon the labour of Ghanaian nurses, and she advocates a more reciprocal relationship with the people and the government of Ghana. Likewise, she praises the then Mayor, Ken Livingstone, for his plans to recognise Londoners' contributions to global warming and his support for the tax justice network as evidence of a more geopolitical understanding of the responsibilities of London's people to the rest of the world. In this vein, Doreen also draws on the innovative performance art of PLATFORM to provide a vivid example of the ways in which London's wealth is dependent upon environmental and human devastation caused in other parts of the world – in this case, by oil extraction in the Niger Delta, West Africa. As Doreen puts it towards the end of the book, this approach 'raises questions of unequal interdependence, mutual constitution, and the possibility of thinking of placed identity not as a claim *to* a place but as the acknowledgement of the responsibilities that inhere in *being placed*' (2007: 216, emphasis in original). As such, Doreen's argument advocates a geographically informed politics that looks outwards to explore the relational construction of people and place in order to embrace the politics of 'place beyond place'.

In many ways, these arguments chime with the concerns of our time. Many political activists are grappling with the challenges of climate change, war, international migration and the injustices associated with globally dynamic corporate supply chains. Indeed, many of the most powerful social movements of the contemporary world speak to Doreen's agenda: Oxfam, Greenpeace, the global justice and tax justice movements are advocating that we take responsibility for our impact on the rest of the world in relation to trade, poverty, war and environmental destruction.

In addition, as indicated at the start of this chapter, there is little doubt that our relationship to place has been changed by the impact of globalisation, increased rates of mobility and entrenched consumer culture over the past 30 years. Evident in the increased ease and incidence of travel and the sharp rise of rates of geographical mobility, many are less rooted in place than they were in the past. While people are still resident in places, their lives, connections and affiliations may be largely non-local (Amin, 2002b, 2006, 2010), and this adds weight to Doreen's plea that we engage with the politics of 'place beyond place'.

For scholars like Doreen and, particularly, her sometime collaborator Ash Amin (2002b, 2010), however, this analysis means that place is no longer seen as a sound foundation for political identifications and their manifestation in political organisation. The bonds between people are argued to be less secure than they were in the past, people's interests are argued to be sharply differentiated by ethnicity, gender and class, and the places in which people are located are understood to be constructed through non-local relations requiring extra-locally oriented activity to secure any change. Many commentators further suggest that local organisation is only possible through the construction of exclusionary, romanticised and/or backward-looking identifications that are likely to have negative consequences for many of those within the area as well as those further afield (Harvey, 1996: Castells, 1997). The lessons of Doreen's earlier work concerning the importance of locally-rooted traditions and institutions as the bedrock of political organisation seems to have been lost. While Featherstone (2008) has made an important argument highlighting the ways in which local politics is itself always constructed through relations across space, or 'place beyond place', thereby suggesting that it is possible to construct a progressive agenda in place through the incorporation of ideas and practices from further afield, he has little to say about the concrete connections that are made through face-to-face relations in place.

As intimated at the beginning of this chapter, however, and in contrast to hegemonic thinking in the discipline, I would argue that geographers are neglecting something important. While our geographical moorings might be time-limited or relatively unstable, we still live in places that provide opportunities for interaction with our neighbours, with the potential to forge a sense of shared interests in relation to place. Geography might be the glue that can bind us together. Furthermore, in the UK at least, our political structures and systems are organised through a multi-scalar architecture that comprises council, region, nation and European federation (Morgan, 2007). We are invited to vote for – and engage with – the people that will represent us in these arenas. Territory remains critical in shaping our geographical imaginations and our political practice. As Crick (1982 [1962]: 18, emphasis added) suggests: 'Politics arises from accepting the fact of the simultaneous existence of different groups, hence different interests and different traditions, within a *territorial unit* under a common rule.'

In their overview of debates about the intersection between geography and democracy, Barnett and Low (2004: 4) have highlighted this gap between the geographical discipline and 'the concerns of political philosophy and democratic theory'. Indeed, they go on to make a powerful argument about the limits of contemporary thinking about politics in geography. While they recognise that the post-structural turn has broadened 'the range of activities understood to be in some sense political ... [it also] carries the risk of jettisoning any concern for the realms in which politics most obviously still goes on' (Barnett and Low, 2004: 7). In the rest of this chapter, I explore this potential agenda for geography, highlighting the contribution of Doreen's work to its advancement. I suggest that there is a danger that the power of Doreen's scholarship in relation to what I have called the politics *of* place may overshadow the need for a renewed politics *in* place, and that we can usefully return to some of her earlier arguments about the importance of geographically differentiated political institutions, traditions and new imaginaries, as outlined below.

Politics and Place

There has been a well-documented decline in the engagement of citizens in the political process in countries like the UK (Pattie, Seyd and Whiteley, 2004; Power enquiry, 2006). Scholars have argued that the decline in social capital associated with the collapse of locally rooted institutions such as trade unions, churches and political parties has caused an inevitable decline in local political life (Putnam, 2000; Pattie, Seyd and Whiteley, 2004). Others have highlighted the growth of individualism associated with consumer culture and its deleterious impact on traditional commitments to party and class (Giddens, 1994). Yet others have suggested that the old politics of locally rooted representative democracy are being replaced by a new interest in social movements and Internet activism (Norris, 2002).

Scholars are largely united in declaring that the old democratic system is now in decline; the structures of democracy remain, but they are agreed to be 'hollowed out' as real power has shifted to other agents and as citizen engagements are focused elsewhere.

In this context, it is important to acknowledge that a number of geographers have highlighted the extent to which there has been a strong tendency towards the *re*-territorialisation of politics in the contemporary world (Brenner, 1999; Dicken, 2004; Morgan, 2007; Castells, 2009; Jones, 2009), as regional governments (both in relation to the city regions and federal alignments such as the EU) have become more rather than less important in political life. However, this said, the energies of most political geographers have been focused elsewhere. Researchers have explored the decline of public space, the spatial extension of the public sphere to the Internet, the political significance of borders and the growth in transnational social movements (Newman and Paasi, 1998; Pickerill, 2003; Routledge and Cumbers, 2009; Staeheli, 2010). Strangely, we have largely ignored the strand of political theory that emphasises the importance of place in the formation of social relationships, the creation of the public realm and the vitality of political life. Writing in the civic republican tradition and following in the wake of Thomas Jefferson and political theorists such as Hannah Arendt and Michael Sandel, Kemmis (1990) argues that there is a concrete connection between the local public sphere and the health of the wider polity. Put simply, he argues that if people do not associate with each other, they are unable to hold their elected representatives to account and political – and social – life is thereby diminished.

Place is particularly important in providing the opportunity for people to forge collective identifications around common interests. Doreen herself has argued that space is 'the sphere of the possibility of the existence of multiplicity in the sense of contemporaneous plurality' and 'co-existing heterogeneity' (Massey, 2005: 9). It is in place that we encounter difference and, as inhabitants, we have to negotiate those differences or suffer the penalties of not doing so by finding ourselves living in uncivil communities with a poor-quality commons (Amin, Massey and Thrift, 2000; Amin, 2006; Boyd, 2006; Fyfe, Bannister and Kearns, 2006; Vertovec, 2007). As Doreen suggests: 'Place … does – as many argue – change us, not through some visceral belonging (some barely changed rootedness, as so many would have it) but through the *practising* of place, the negotiation of intersecting trajectories; place as an arena where negotiation is forced upon us' (2005: 154). In what she has also called the 'throwntogetherness' that occurs in place, we are forced to construct a public realm to find ways to live with each other. Geographers are largely agreed that such politics will be shaped by divergent collective identifications – although the focus has been on social movements rather than local democracy – and conflict as well as consensus (Massey, 1995a; see also Purcell, 2008).

Moreover, it is important to recognise the spatially uneven cultural and political inheritance that will shape this politics in place. As we have seen, Doreen's earlier work documented the uneven development of political infrastructure, traditions and practices across the UK (Massey, 1991a). Such legacies can act as stimulants as well as impediments to political experiment, and in the USA, for example, a radical new labour movement has emerged in precisely those areas where the traditional trade union movement was weakest (Milkman, 2000). Cities like Los Angeles have developed new forms of community unionism, moving beyond the traditional focus on workplace trade union organisation more easily than the labour heartlands of the industrial North (Soja, 2010). Likewise, some of the most powerful innovations in relation to labour organising in the UK – such as London Living Wage campaign – have been driven by those outside the trade union movement, as outlined at the start of this chapter in relation to Citizens UK (Wills, 2004, 2009b; Wills et al., 2009). The speed at which political activists can reinvent tradition is thus sometimes more significant than their ability to sustain tradition through difficult times, and this, in turn, depends on the people living in any particular place.

My recent political experience working with the broad-based community organisation Citizens UK, outlined at the start of this chapter, indicates the potential role of place in providing a platform for forging collective identifications, shared interests and political organisation. Its local alliance, London Citizens, has found the islands of organised civil society – faith, educational organisations and trade union branches – that generate social capital in communities and has then created a super-ordinate identity category that links them together (Jamoul and Wills, 2008; Wills, 2010, 2012). Over time, the connections between local leaders in these organisations become stronger even as the members of their church, university department or school are constantly changing. As Nicholls (2008: 842) writes of urban social movements more generally, place can act as a 'relational incubator' due to the density of leaders and the opportunities that exist to develop long-term relationships across difference.

I would argue that the time is ripe for a renewal of research into the relationships between place and politics. While the democratic deficit occupies popular and academic imaginaries, there is important work to be done in exploring the relationship between local people and the democratic process, documenting the ways in which local communities can and do organise themselves to challenge the political order. Moreover, there is now scope for a normative intervention in relation to the connections between place and politics. Echoing the work of Gibson-Graham (2002, 2004, 2006, 2008) in relation to the generative power of language in rethinking economy and the role of place as a site for political re-subjectification, geographers could make a much stronger intervention in relation to the necessary connections between place, the public and politics (see also Barnett and Low, 2004).

Doreen provides important tools in theorising the challenges involved in this work. An integration of the insights from her long career, encompassing what I have called the 'politics *in* place' and the 'politics *of* place', would provide solid ground from which to begin. Going back to the example of Citizens UK, the organisation seeks to build relationships between people in particular places as the foundation for securing the power to act and shape the local community. In so doing, local alliances reflect the institutional inheritance and affiliations of the local area. In then acting for change, such alliances will also expose the networks of power relations that make that place what it is, necessarily challenging vested interests in pursuit of improvements in life. Organising diverse local communities can provide the motor for change, remaking place in the process.

Writing in the wake of the general election in the UK, when all the politicians promised to transfer 'power to the people', geographers and others can build on Doreen's intellectual and practical legacy in analysing and contributing to the ongoing process of democratic reform. Powerful communities will remake their places through politics, and in so doing remake their relations with the rest of the world.

Note

1   The full title of this programme was *The Changing Urban & Regional System in the UK* (CURS) but it was popularly known as the Localities programme and included research in Birmingham, Cheltenham, Lancaster, Liverpool, Middlesborough, Thanet and Swindon (for more on the history of this work, see Cooke, 2006).

Chapter Ten
A Global Sense of Place and Multi-territoriality: Notes for Dialogue from a 'Peripheral' Point of View[1]

Rogério Haesbaert

I first met Doreen Massey when I stepped off the train in Milton Keynes on the way to the Open University, to begin a post-doctorate under her supervision. She, with her special blend of ironic tenderness, opened the conversation by saying, 'So, you are real …' Until then our contacts had been mediated by the Internet or friends such as Felix Driver and Luciana Martins. We then began to mark our paths on the crossroads of trajectories (a 'space', in Doreen's terms) never before imagined, passages that would extend from cold afternoon coffees at the British Library in London to hot lunches on sun-drenched beaches in north-east Brazil.

During all these encounters, what I learned to admire most in Doreen was her ability to move, without difficulty, from a more or less arid (yet never politically disassociated) theoretical discussion to plain yet profoundly human commentaries about such things as the details of a landscape (during train trips to Milton Keynes), the brilliance of a star (in the night sky over the town of Jericoacoara in Brazil), the semi-arid qualities of the *caatinga* (in the backlands of north-east Brazil) or, simply, the way a spider wove its web on a winter morning at the Open University campus. Far from being one of those affected and aloof intellectuals, Doreen seemed to have remained loyal to her working-class upbringing on the outskirts of industrial Manchester. She always kept her feet on the ground; in debate, she would defend her ideas unrelentingly and yet returned easily to 'common sense' examples, to the simpler sentiments that intuitively enrich our daily lives.

Spatial Politics: Essays for Doreen Massey, First Edition.
Edited by David Featherstone and Joe Painter.
© 2013 John Wiley & Sons, Ltd. Published 2013 by John Wiley & Sons, Ltd.

One time, I made an assertion that she greatly appreciated. We don't always have to be concerned to exhaustively dissect a subject, I said, citing all the authors that contributed to an idea (something that appears to be an obsession among many Anglo-Saxon authors). Rather than affirm our eruditeness, I continued, we should seek to produce innovative work, a little like a good novel, where the author invests all their forces and originality. A great geographer, therefore, is not the one who can demonstrate dominance of the literature in his or her specialty, but one who is able to contribute something new, to offer an innovative reading and thus make a unique contribution to their whole field.

Doreen is a great geographer because she knows how to read the world with her own eyes, bringing perspectives never before developed, combining broad theoretical-philosophical reflection with an intensely rich intuition. (Could that be why she appreciates – though in a decidedly critical manner – an author like Henri Bergson?) One could call this state of awareness 'geographic intuition', a way of looking at the world through space – something many claim but few truly possess. Fernand Braudel has defined this same quality of a geographer's gaze with a simple and direct phrase: 'the study of society through space' (Braudel cited by Baker, 2003: 22).

Doreen is also a great geographer because she constantly seeks to overcome the discipline's large dichotomies: materialism–idealism (though she leans towards the first, she distances herself from structural economism), objective–subjective (while often belittling the former, she regularly emphasises the latter), nature–society (despite much greater attention given to the second, she has explicitly examined their relational sense, as demonstrated by an 'addendum' on the concept of place in her work *For Space*).[2]

Her idea of the 'global sense of place' is one of her principal, universally recognised contributions to the geographic debate, one that none doubt, so great is the quantity of citations (even if it is to argue with the idea as, for example, Negri and Hardt do in *Empire*, 2001),[3] and it is this idea that I want to emphasise here. I propose to briefly reread her work in light of a conceptual proposition to link her 'global sense of place' to processes I refer to as the construction of 'multi-' or 'trans-' territoriality, even though the idea obviously deserves further development.

Place is to Anglo-Saxon geography what territory is to 'Latin geographies'.[4] Sometimes the words change but the concepts they carry remain very similar.[5] To be faithful to the political commitments so emphasised by Doreen, it is important to focus attention on the questions these terms address and the problems they seek to resolve, rather than the words themselves.

During the last few decades, Latin America, and Brazil in particular, have experienced a real avalanche of political manifestations involving, directly or indirectly, territorial questions. At present, the region is one of the richest spatial contexts in terms of struggles seeking social transformation in the form of cultural recognition and/or the redistribution of wealth – especially redistribution of land, in the reconstitution of the rights of landless and homeless groups as well as 'minorities' (of Indigenous peoples, for example, who in some countries are clearly the majority). These are processes that occasionally manage to reconfigure states – as Doreen herself has seen through her experiences with the Sandinista government of Nicaragua during the 1980s and the Venezuelan government at present.

Territory in this discussion is not just a 'state question'. In Latin America today, I can affirm, getting (re)territorialised is a political strategy of transformation much more than an academic question; it is a lived, practised and practically 'demanding' question. Some might consider this emphasis on territory exaggerated, but many of these struggles are overwhelmingly territorial struggles. Further, this positions territories not in an abstract sense of simple formal recognition within the territorial sphere of the state, apart from the heterogeneity of the experiences of social groups, but as initiated through particular practices, dilemmas and meanings. We could say, in fact, that these struggles/social practices themselves continually remake the concept of territory.

Place, especially from the perspective of Doreen Massey's cosmopolitan London, can be defined as 'processes' that 'do not have to have boundaries, in the sense of divisions which frame simple enclosures'. They also 'do not have single, unique "identities"; they are full of internal conflicts' (Massey, 1994a: 155). Still more important, the specificity or 'uniqueness' of place, fundamental to her definition of the concept, derives from the condition that 'each place is the focus of a distinct *mixture* of wider and more local social relations' (1994a: 156, emphasis in the original), and its singularity comes from a specific combination of interactions at multiple levels/ networks.

In a more general sense (one inspired by Gilles Deleuze and Felix Guattari), it is interesting to consider that concepts, such as those examined here, are viewed more as 'transformers' (Holland, 1996) than as 'revealers'. Thus, concepts function as producers just as much as products: 'the concept is exactly that which gets us thinking. If the concept is product, it is also producer: producer of new thoughts, producer of new concepts and, above all, producer of events, to the extent that it is the concept that frames the event, that makes it possible' (Gallo, 2003: 43).

In contrast with 'science' (as defined by Deleuze and Guattari, who make no explicit distinction between natural and social sciences), which seeks to specify and stabilise specific domains of reality, philosophical concepts intervene in problems to cause disruption, thereby creating new connections with other concepts as well as a given historical and geographical context. According to Holland (1996), the formal definition of a concept is often less important than knowing how to deal with it, how it 'works' and what can be 'done' with it. Thus, concepts 'do not possess independent content, something autonomous from that which they acquire through use in a given context' (1996: 240).

It is in this sense that 'territory' in Latin American geography comes close to 'place' in Anglophone geography, because there are deep similarities between 'the content they acquire through usage' in each context. Thus, territory in Latin America, just as place in the conceptualisation of Doreen Massey, can be approached as:

1   *'process,'* in a relational sense, because, in addition to emphasising the dynamic, action qualities of territorialisation (de- and re-territorialisation) more than its characteristics as a 'product' (territory 'in itself'), we recognise territorialisation for its mobility. Inspired by Deleuze and Guattari, we define produced-territorialisation by the repetition of movement, a movement somehow under control. In this light, clearly, the majority of nomadic peoples do not represent a model of deterritorialisation, as many have argued (Haesbaert, 2004). This conception of 'relational territory' is in contradistinction to simplistic analysis like those of Amin (2004) who, in proposing a 'non-territorial reading of a politics of place', argues for the opposition of 'territorial' and 'relational' reasoning – as if 'territory', restricted to a now traditional approach, were reduced to well-circumscribed legal-political limits.

2   *'frontiers'* or *'borders'*. These too need not be well defined nor even marked by a clear distinction between who's 'in' and who's 'out' (even though borders most often are more political-territorially definable than the delimitations indicated in the conceptualisation of place), as ambiguous situations proliferate these days, even in those territories apparently better marked (such as the *camps* of 'inclusive exclusion' discussed by Giorgio Agamben, 2002, territories of 'legal illegality' at once inside and outside state jurisdiction).

3   *multiple identities*. In this case, territory has its conceptual 'focus'[6] in power relations – that is to say, in multiple powers, especially in that which can be referred to as the intersection of different scales and modalities of power. It is in this sense that we propose to conceive of territory from the perspective of multiple overlapping power relations, from the most obvious material processes of political and economic power to more subtle – yet no less significant – processes of symbolic power (Bourdieu, 1989).

Obviously we are dealing here not with isolated concepts, but with 'constellations of concepts' (Deleuze and Guattari, 1992), because they work only in relationship to other concepts in a larger assemblage. We can also speak of 'hybrid concepts' or of the 'hybrid nature' of the concepts (Santos, 1996), to the extent that we not only express this heterogeneity inherent to the concept, but also propose others, concepts that express within their very etymology ('network-territory', for instance) this mixture, these passages. In these ways, important relationships can be established between our concept of 'multi-' or 'trans-' territoriality and Doreen Massey's 'global sense of place'.

Multiplicity is a marked characteristic in Massey's 'global' and politically 'progressive' conception of place, as well as our relational and open conception of a multi- or trans-territoriality, from Brazilian hybrid spaces (Haesbaert, 2004, 2011). For Massey, the fundamental characteristics not only of place but also of space in a more general sense are defined by multiplicity: 'we understand space as the possibility of the existence of multiplicity in the sense of contemporary plurality; ... the sphere of coexisting heterogeneity' (Massey, 2005: 9).

In reproducing Bergson's distinction between differences discrete (related to extended magnitudes and distinct entities) and continuous (intensity, fusion) – two of the fundamental forms by which multiplicity is made manifest – Massey profoundly criticises his definition of the relationship between space and discrete differences, that restrict spatiality to the countable characteristics of the real. For Bergson (1993 [1927]), the effectively 'multiple' is referred to only as a temporal dimension, because it corresponds to a qualitatively 'real' (or intensive) differentiation. In the interpretation of Massey, obviously, space incorporates, concomitantly, the sphere of the multiple, both in the sense of its quantitative accountability (so prized by the capitalist order, we could say) and qualitative change.

This is one of the ways that we refer to 'multi-territoriality' – recognising, in the first place, the simultaneous existence of a 'multiplicity of territories' (different types or species of 'extensive' territories) as well as *the* 'multiplicity of territory' (territories, in and of themselves, characterised by strong internal differentiation or intensive, continuous multiplicity).[7] In both cases, one can argue that it is possible to articulate a multi-territoriality – or, if you will, construct a 'global sense of place', potentially amplifying or making more explicit the complexity of Doreen Massey's conception.

It is important to emphasise that multi-territoriality has both broad and strict meanings, the latter linked to processes of globalisation to which we are presently subordinated. In the broader sense of 'enouncing a multiple territoriality', multiterritoriality 'is not exactly a novelty, for the simple fact that, if the territorialisation process takes off from the level of individuals or small groups, every social relationship implies territorial interaction, an intersection of different territories. In a certain sense, we would have always lived a 'multiterritoriality' (Haesbaert, 2004: 344).

One of Massey's central references has been her own neighbourhood in London – a kind of 'multiple territory' (or place), with its cultural diversity connecting multiple networks from various corners of the globe, forming a kind of 'network-place'. We can affirm our ability to construct places in/through mobility in the sense of articulating among distinct places/ territories, however, not only simultaneously but successively through stressing 'temporal multiplicity' as well as 'spatial multiplicity'. These possibilities are especially marked when we consider the overall power-geometries of our different mobilities, as Doreen would say.

We thus distinguish living a *simultaneous* multi-territoriality – in its symbolic as well as functional dimensions – when we experience (and, thanks to virtual reality, often actually control) diverse territories at the same time, without the necessity of physical mobility – and *successive* multi-territoriality when we share distinct territories and, through mobility, network among them (creating another kind of network-territory). In the latter case, multi-territoriality is constructed through the physical articulation of various territories; in the former, multi-territoriality derives from the intersection of multiple networks, without requiring us to intensify our physical mobility.

Perhaps more intensively than in Massey's global place, we can 'control' different territories at a distance, complicating even more the construction process of our multi-territoriality. Informational control, articulated at long distance, is one of the ways our multiple identities with place – or our ability to retain (multiple) control of territories – will certainly intensify in the years to come. Today, we are not only able to conjoin our place with alien places, we can also exercise diverse forms of territorial control at a distance without physical mobility. This could be a process of multiple territorialisation accomplished with a minimum of physical movement.

While interesting, these references – simultaneous and successive multi-territoriality – are not stagnant distinctions, because real situations – in varied intensities – always involve these two facets, especially if we consider that symbolic power is also involved in deterritorialisation processes. The terms help deepen our understanding of multi-territoriality in symbolic as well as functional ways. Thus there would be (multi-) territorialities of a higher functional sense and others of a higher symbolic sense, according to the force of functions and meanings they are given by different social subjects in (territorialis)action.

Producing hybrid cultures, for instance, can also means building somehow hybrid, threshold or 'trans-frontier' territories, multiple territories whose manifestation interferes directly in our conceptualisations of the world, in the construction of our social identities. Contemporary multi-territoriality itself can therefore facilitate processes of hybridisation, either by increasing our physical mobility, articulating more than one territory (as occurs with migrants in a diaspora), or by the in situ territorial diversity itself, as occurs especially in the cosmopolitanism of districts inside large global cities. In a successive or wider sense of multi-territoriality, our hybridisation is, say, induced by mobility and 'control'. For this successive multi-territoriality it is essential that we experience physical displacement in order to have our multi- or trans-territorial experience – with the important caveat that, of course, not everyone moving through different spaces necessarily lives a multi-/trans-territoriality beyond its purely functional character.

To the same extent that deterritorialisation is not, in a relational approach, synonymous with mobility (as we can be profoundly 'deterritorialised' in a prison, for example, by the fact that we are no longer the ones exercising control over our territory),[8] additional physical mobility or territorial openings do not mean that we will automatically be experiencing an intensification of our multi-territoriality. Under these circumstances, we have to distinguish between potential and actual multi-territoriality.

An effective and 'global' multi-territoriality in the successive sense is not necessarily accomplished by one's circulation through more than one territory and its articulation in a network, as this could occur in a merely functional way. Think, for example, of globetrotters or business executives who travel frequently, circulating on a global scale, but tend to stay in the same sorts of spaces (hotels, stores, restaurants), generally avoiding interaction with the great cultural diversity of the territories they visit. Here and there they may even cross paths with Others, but it is as if 'Others' had been rendered invisible, without dialogue – or, when a mandatory dialogue takes place (as in hotels/restaurants and shopping situations), it is a contact of an almost purely functional character.

The same can occur in the case of 'global' openings of our territories and places: it is not simply the fact that a neighbourhood is open to a great diversity, to cultural hybridism, that makes its inhabitants take part in this multiplicity, enjoying the intensity of multi-territoriality (or, if one prefers, *trans-territoriality*, to emphasise in-between or transit across diverse territorialities). Exactly the opposite effect could occur: faced by growing diversity, as individuals or small groups we might recoil to the relative seclusion of familiar safe territories (or 'places').

As indicated initially, we cannot forget that a form of territorialisation also takes place 'in and through movement'. Many are those who, nowadays, identify themselves with this mobility so that for them territory, like their identities, is built by the amalgamation of multiple territorialities, or more 'radically', by 'moving' or 'transiting across multiple territories', which also leads us to think about another kind of trans-territoriality.

In affirming that 'there are no rules of space and place', Doreen Massey (2005: 163–176) argues that it is not the closure or openness of places/territories that establishes behaviours and social meanings, creating a species of spatial fetishism, but its participation as indissociable constituent of these relations and one that is profoundly engaged in multiple transformation possibilities.[9] Although it has not been Massey's objective to extend the debate about the multiplicity of space (in the sense of its multiple trajectories/networks – a network logic – as well as of its multiple extensions/areas – a zonal logic) to the concomitant multiplicity of time, she has discussed the 'multiplicity of the spatial' as a 'precondition for the temporal'. She further adds that 'the multiplicities of the two together can be a condition for the openness of the future' (Massey, 2005: 89).

It is this differential 'historic density' (or 'unequal accumulation of time,' in the more materialist analysis offered by Santos, 1978) of profoundly diverse multiple temporalities that transit Latin American territories and make us rethink a series of propositions – such as 'a global sense of place' or multi-territoriality – in light of our own experiences and contexts. Latin America – and Brazil – are today true laboratories of new cultural and political experiences, spaces marked by resistance, precursors of post-colonialism (dating from the nineteenth century, the largest, most diverse [almost] continental space to formally break with the colonial system) and of hybridisms (found in the unusual practices of cultural encounter, such as those of Amerindian-European-African, as well as in forms of thought, such as the anthropophagic philosophy of Brazilian writer Oswald de Andrade).[10]

An early lesson that we learn in a 'peripheral' context like Latin America is that deterritorialisation, which is neither good nor bad in and of itself, applies more often to the subalterns who predominate on the margins of the world economic order than to the minority of 'global citizens' who circulate freely around the world.[11] Moreover, deterritorialisation as 'exit of a territory' does not necessarily imply a better and more positive reterritori-alisation – the 'entrance' of a new territory – tending instead to correspond, in the experience of many subaltern groups, to a process of increasingly precarious territorial construction.[12]

(Re)territorialisation processes involving social groups supposedly defensive of closed, more stable and conservative territories and/or places are being forged today in Latin America and especially Brazil. In the case of Brazil, these groups were officially designated as 'traditional peoples' (a controversial term formalised in Brazilian law), but this does not mean that they plan to construct 'traditional' territories/places. We have to be careful, in this case, to avoid seeing these processes as binary approaches between conceptions of territory and/or place that are either 'traditional' *or* '(post) modern,' 'conservative' *or* 'progressive'. These Brazilian examples are quite representative of the ambivalence with which these properties are constructed. In these ways, too, territories become 'multiple' – manifesting a multiplicity of identitary and power situations – identity and power that alter themselves or mix together depending on such factors as the historical and geographical context (in terms of action scales) in which they are constituted.

Territorial closure (always relative) or more acute territorial delimitations do not necessarily mean the defence of a retrograde, conservative political vision. It can represent, as the so-called 'traditional peoples' of Brazil demonstrate, a moment inside a greater struggle and one that does not dichotomise 'tradition' and '(post)modernity', but merges them, bonded together, in a new post- (or de-)colonial amalgam. For 'traditional peoples' like *indígenas* (Indians) and *quilombolas*,[13] at the moment of the clear establishment of the physical limits of their 'reservations', relative territorial closure can result in creating the very conditions necessary for the survival of the group as such. In the case of original *quilombos*, the isolated and clandestine qualities of these practically closed-off runaway communities were synonymous with freedom.

While these 'reborn territories' may have resulted from a violent process of deterritorialisation or precarious territorialisation – the 'rest' of what was left over from a predatory process of conquest – these groups may find in the reservation territories conditions essential for their reproduction as culturally distinct groups. Moreover, these 'reborn territories' can represent a clear demonstration of the subversion of the majority order – imposing, for example, common use of the land, a distinct relationship with nature – and thus become spaces for avoiding strictly capitalist processes of incorporation in the form of land exchange value.

In the same way, the openness of these territories to external articulations – global connections, for example – does not automatically translate into a progressive attitude. Large transnational firms (these, too, though in other ways, engaged in 'multi-territorial' or 'trans-territorial' processes) and even some non-governmental organisations have their eyes on the resources of these territories, on the biodiversity of the areas that, exactly because of becoming 'reserves', guarantee the possibility of future exploitation of their multiple biogenetic patrimony.

Thus, if (theoretically) 'there are no rules of space and place', as Massey argues, political (or politico-economic) rules exist that guide a great number of these actions. While submerged in profoundly unequal power-geometries, the hidden rules are generally controlled by very well-territorialised groups (in 'network-territories' with well-defined circuits) and their 'reservations'[14] clearly guaranteed on the world map. The intensification of the globality of our 'senses of place' does not in itself mean something either positive or negative. Even though politically we could – and probably should – defend a greater 'global sense of place', an intensification of multi-territoriality, attributing to spaces, in this way, less exclusivity and closure, obviously it is not simply openness that will guarantee a distinct and more just political condition.

There is an immense and intricate game of openness and closure in the current global – or, perhaps, to be more precise, 'glocal' – situation. If 'seclusion' – in the Foucauldian disciplinary sense – no longer makes sense or is in profound crisis, 'populations' (in the biopolitical sense proposed by Foucault) continue to be, not exactly 'confined' or 'enclosed', but 'contained', through processes that we denominate as 'territorial contention' (Haesbaert, 2008). There, new walls and fences acquire the effect not exactly of confinement – through definition of areas – but of contention – as barriers to networks, especially, today, the flux of people. Due to this simple barring effect, the flows end up searching for other spaces to go – as demonstrated by many current migratory flows. These factors reveal a much more complex process of territorial permeability and juxtaposition.

The de/reterritorialisation movement of so-called traditional people in Brazil today, much more than the simple construction of well-bounded and stable territories, is in fact a way of recognising that, in its variable multi- (or even trans-) territorial character, this process is always present and remade. De/reterritorialisation is also, at different levels, a type of 'being in-between' or an activation/production process of distinct territorialities – that means entering into a game of multiple identitary situations and multiple relations of power. To have consciousness of this multiplicity and know how to 'play' with the diversity of situations generated by de/reterritorialisation processes is strategically fundamental to the political action of these groups.[15]

What matters here is the openness of territories to continually extend possibilities to enter, leave and/or pass by, creating conditions for our insertion, if necessary, in the territory of 'others', places that then pass through an ambivalent experience of being 'ours'. As Massey writes: it is 'the event of place ... in the simple sense of the coming together of ... previously unrelated [trajectories], a constellation of processes rather than a thing ... Place [territory, in this case] as open and as internally multiple' (2005: 141).

Massey (2005) returns to emphasise how these characteristics of openness and event that place implies also have political implications. In fact, they are rather political, particularly in light of a Deleuzean approach to concepts. Many of our practices – and our new walls, fences and 'camps' ('territories of exception', through the language of Agamben, like refugee and immigration control camps) – although ambivalent in terms of in/out relations, wind up representing more the inverse process, as they make more difficult, in one way or another, contingency and chance encounters.

More than simply defending the existence of a dense multi-/trans-territoriality – or, in a similar way, of a general 'global sense of place' – it is important to be cognisant that, even though some may be more closed, others more open, all territories should be constructed from a base of respect for the multiplicity of space and the life that animates it. Doreen Massey, through her intellectual trajectory, has always sought to promote this respect and commitment. For Massey, 'science' is constructed only when engaged/implicated in the movement for an effective transformation of society – even if this transformation is always in the form of a reaffirmation of an openness to re-evaluation and change, of a permanent sense of becoming. We should not forget, however, that this openness to change involves a difficult game of mobility and rest, as no one should let themselves be carried away by either a 'dictatorship of enclosure' (ghettoification) or its opposite, the 'dictatorship of movement' (as Paul Virilio would say). In the words of Cornelius Castoriadis:

a subject is nothing unless it is the creation of a world for-itself in a relative closure ... This creation is always creation of a multiplicity ... This multiplicity is always deployed in two modes: the mode of the simply different, as difference, repetition ...; and the mode of the other, as otherness, emergence, creative, imaginary, or poietic multiplicity. (Castoriadis, 1997: 375)

In a supposedly 'in-secure' society such as ours, one increasingly shaped by differences/inequalities and based on situations decidedly more ambiguous than worn-out notions of insiders and outsiders, openness and closure, we end up denying ourselves this effective otherness. Otherness instigated by the permanent criss-crossing of trajectories (as Doreen would say), through spaces strategically open (or 'relatively closed', to return to Castoriadis), is the true field of battle to join the struggle for equality and confront differences in the event of encounters, condition sine qua non for the emergence of the effectively new.

Acknowledgements

The author wishes to thank Clifford Andrew Welch for the translation, Dave Featherstone for suggestions and substantial improvements of his English, J. Simon Hutta for his comments on the earlier draft of this article and Arun Saldanha for his encouragement.

Notes

1 Here I use 'peripheral', with inverted commas, to mean not just that the 'centre is also in the periphery', but mainly in the sense that through 'peripheral' spaces we can produce our own (and new) centralities.
2 In this case, as exemplified by Bruno Latour, by means of expanding the universe of the 'non-humans' as part of the fundamental characteristics of space and place: 'the nonhuman has its trajectories also and the event of place demands, no less than with the human, a politics of negotiation' (Massey, 2005: 160).
3 A response she refutes in *For Space*, arguing that their allegation depends on a false dichotomy between 'the romance of bounded place and the romance of free flow' (Massey, 2005: 175).
4 Geographies in Spanish, Portuguese, French and Italian languages.
5 Though polemical, we find in English-language literature, such as that of the philosopher J. E. Malpas, the joint usage of 'place' and 'territory'. The author affirms that his book *Place and Experience* 'attempted to uncover ..., not an "empty space", but a single complex and interconnected territory – the very territory in which our own identity is grounded and in which the encounter with other persons and with the things of the world is possible' (Malpas, 1999: 195). The philosopher here recalls the frequent but controversial distinction between a space, defined in the abstract or functionally, and a territory effectively constituted by the practices of different social forces.

6   The idea of a conceptual focus (*foco*) is interesting because it helps us to recognise that concepts allow us to concentrate our analytical sights on certain relationships while, at the same time, leave others 'out of focus' without negating or ignoring their presence.

7   Romance languages like Portuguese also facilitate a distinct play on words: 'múltiplos territórios' and 'territórios múltiplos'.

8   In this respect, see my reflections in Haesbaert (2004), especially the chapter 'Deterritorialization in Im-Mobility' (pp. 251–278).

9   'What is certain,' writes Massey (2005: 164–166), 'is that there are no general spatial principles here, for they can always be countered by political arguments from contrasting cases. ... The question cannot be whether demarcation (boundary building) is simply good or bad ... The argument about openness/ closure, in other words, should not be posed in terms of abstract spatial forms but in terms of the social relations through which the spaces, and that openness and closure, are constructed; the ever-mobile power-geometries of space-time.' Here we can insert an observation that, in a similar way, deterritorialisation 'implies identifying and placing a priority on subjects involved in the de-re-territorialization process, submerged in multiple webs that permanently merge distinct points of view and actions that promote that which we can call deterritorialized territorializations and reterritorialized deterritorializations' (Haesbaert, 2004: 259).

10   For an introductory geographic appropriation of anthropophagic discourse, see Haesbaert (2011).

11   For fuller and more critical treatment of the topic, a form of interpreting the deterritorialisation process common to intellectuals from the 'centre', see my 2004 book *O mito da desterritorialização* (*The Myth of Deterritorialisation*, as it is usually seen), published in Spanish (2011) as *El mito de la desterritorialización* by Siglo Veintiuno Editores, México.

12   In Portuguese we use the term 'precarização' meaning the process, or tendency, of becoming precarious.

13   *Quilombolas* are the descendants of formerly enslaved Africans and blacks who escaped slavery to form relatively isolated 'territories of freedom' called *quilombos*.

14   A play on words that exploits the multiple meanings of 'reservation' in two senses, that of booking a seat or reserving a place and that of a designated place/ territory, such as an Indian 'reservation'.

15   The importance of these debates can been seen in countries like Bolivia, where so much energy has been invested in discussing innovative political-territorial conceptions of administration that the term 'trans-territoriality' has been incorporated in legislation.

Chapter Eleven
A Massey Muse

Wendy Harcourt, Alice Brooke Wilson, Arturo Escobar and Dianne Rocheleau

Prologue

Doreen Massey was one of the most direct and influential inspirations for three of us (Wendy Harcourt, Dianne Rocheleau and Arturo Escobar) when, a number of years ago, we started out on a collective project entitled Women and the Politics of Place (WPP). We were, and still are, interested as much in justice as in difference. We set out to explore, with a group of activists and academics, the potential of place as a site of progressive politics vis-à-vis neoliberal globalisation, and the central role of women in it. Reading Doreen's work on space and place in this context spoke to us powerfully as individuals committed to a better world, as feminists, as intellectuals and as academics. Her work was a guide as we moved along with the project. As she says, rethinking a politics of place is 'an emotionally charged issue' (2004: 6). That her work spoke to us emotionally and analytically is no surprise; when a graduate student (Alice Brooke Wilson) also had this powerful response to Massey, we invited her to join this 'Massey Muse'.

This chapter is organised around short statements by each of us on our particular engagement with Doreen's work and its importance for our respective works; each statement is followed by questions that emerged during the collective part of the writing process. We would like to add that we are not writing this piece as experts on her work; rather, we write from a selective engagement with some of her works and, above all, in the spirit of

Spatial Politics: Essays for Doreen Massey, First Edition.
Edited by David Featherstone and Joe Painter.
© 2013 John Wiley & Sons, Ltd. Published 2013 by John Wiley & Sons, Ltd.

care and responsibility that her scholarship so wonderfully displays. The engagement with her work in these pages is thus an act of care as much as an attempt at a theoretical and political discourse.

Wendy Harcourt: Space, Place and the Body Politics of Development

I was introduced to Doreen Massey's work by Arturo Escobar as we were setting up our own examination of Women and the Politics of Place (WPP) in 2001. We were interested in how women's struggles around the defence of place offer sources of creativity, culture and alternative development, and how these place-based activities interacted with global processes (Harcourt and Escobar, 2002, 2005). In a written contribution to our project published in the journal *Development*, Massey (2002) warned us to not romanticise place as we were defining it around the body, community, home, environment and public political engagement. She was wary of us grounding international questions in what we perceived as meaningful local contestations of place. She warned that the place 'closest in' might not always be the first priority. Space is concrete and embedded too and should not be seen as abstract and in this way cede space to the powers we are contesting. She asked that we embed our politics of place in the politics of space rather than rush to give priority to place above all.

Massey's early work underlines the complexity of considering gender 'in place', showing us how the 'intersections and mutual influences of "geography" and "gender" are deep and multifarious. Each is, in profound ways, implicated in the construction of the other: geography in its various guises influences the cultural formation of particular genders and gender relations; gender has been deeply influential in the production of "the geographical"' (Massey, 1994a: 177). She taught us that gender, not just capital, affects people's experience of place and space. The WPP project echoed these concerns, finding that 'patriarchy varies from place to place depending on the power plays, but it is always present and, in using the term, we are underlining the constantly unequal relations of power between men and women as well as among women and among men' (Harcourt and Escobar, 2005: 4).

As I continue to use our Women and the Politics of Place framework in both my writing and activism I find myself returning to Massey's insights. Feminist resistance (in both activity and theory) to globalisation tends to be spoken about at the micro level. There are many case studies of women 'in place', contextualised and graphic, depicting the micro realities of oppression and resistance. But the bigger macro picture is left disembodied and non-gendered, or spoken about in abstractions. Massey's later work, showing how identity and place construction fit in a post-structural feminist understanding of power relations, provides a key theoretical tool for pushing to 'widen' our understanding of spatialisation to see the potential for 'not merely defending the local against the global but seeking to alter the very mechanisms of the global itself ... a local politics *on* the global' (Massey, 2004: 11, emphasis in original).

To take one example in mainstream development policy, the work in which I am mostly engaged: microcredit has been proposed by the World Bank and civil society groups as the way that poor women can best engage in economic development. There are many studies and popular pictures in the international press of women in Bangladesh, for example, with mobile telephones building tiny autonomous enterprises with loans from the Grameen Bank. The pictures have as a backdrop rural villages, family and community with the women smiling into the camera, becoming 'empowered'. They are the symbol of successful 'gender and development' by bringing the market system to the village and making women agents of their own empowerment. Presented as good practice, this banking 'in place', led by once victimised women, becomes a success story of development.

In contrast, the economic space that actually defines development policy is the place of abstraction. As a team member of a major development report, I was recently encouraged to bring the 'micro picture' of women's lives to the report. This meant that the strategy of microcredit and how it can work for women could be set out in a tidy box in the report. But in the core analysis of the larger structures affecting women's lives, nothing connects. The abstractions remain very hard to get through.

The impact of the current global financial crisis on precisely those poor women is obscured by the crashing numbers, failing stock markets and falling governments. There is little sense of how such a crisis is grounded in people's lives; discussion is played out in the formulas and calculations of expert economists. Place-based analysis of women's and men's lives 'on the ground' is seen as peripheral to the real issues being set up by experts in universities, world banks, governments and UN systems – not that we get smiling pictures of expert economists sitting around tables with their PCs and PowerPoints, sipping bottled water and snacks, undoubtedly empowered by the development process. How these people and their institutions are made up of connections and relations is not open to scrutiny.

The relation and connection of these empowered development analysts to the absent others is indeed an interesting question. What are the responsibilities to the rural women styled as microcredit entrepreneurs by the people sponsoring and writing the report? To place them in a box in a report, while the main decisions are taken by 'experts'? Trying to engage in these debates, I am reminded of Massey's warnings. There continues to be a dangerously dismissive romanticising of place. Those who have lived in poor countries and can describe the difficulties of place are welcomed to give their stories as interesting boxes to the main text. But the main storyline is the abstract description of economies and government policies done in academe and policy bureaucracies far from the places and people who are the subject of the report. *Who* are doing it are European men (mostly) and some women who are paid hundreds of times more for their brief engagement in the project than those for whom they are prescribing solutions will ever see in their lifetime. This politics is not questioned.

Crucially, those who work 'on the ground' are no less romanticising in their essays and reports, which decry or ignore the abstract figures and numbers of mainstream reports. The many stories of alternatives to development in social movements (including the feminist movement), speaking to their own sense of place, are as often disembedded as mainstream reports are disembodied. As Massey warns, too often these stories fail to connect to the economies and politics which structure their descriptions. I concur with those who hold a negative assessment of microcredit as a strategy to bring women into development. They are right that eulogies of microcredit fail to take into account the multifarious structures which are determining and limiting women in poor rural communities from changing their lives. But too much focus on place, as in the feminist case on the repression of embodied place (with the body as the place 'closest in'), leads to its own blindness.

To take another example of how a focus on the place closest in can lead to counterproductive outcomes: I attended, in November 2008, a global feminist movement meeting in South Africa. The capacity-building sessions on how to record on video and iPod personal stories of sexual gender-based violence were full, but the analyses of financial crisis were confined to the corners of the event. In other words, the conference failed to address the looming financial crisis and the larger structural constraints on women's lives and politics.

As Massey states, it is critical not to counterpose place and space. As my examples show, it is in theory and reality difficult to bring these ways of seeing together. What I have learnt from Massey is to not be content with the divisions between 'micro' and 'macro', 'personal' and 'political', 'local' and 'global', or any attempt to say, 'what is real is here at home, what I feel', in some innocent sense of what structures reality. I continue to look for the ways place and space shape interactions in struggles for justice.

Questions for Wendy

I really like your image of seemingly disembodied (mostly) groups of white men and some women in 'serious' meetings discussing embodied/embedded women, who end up as romanticised images in 'boxes' in technical reports. In your book, Body Politics in Development *(2009), you take up these issues, by asking a series of complex and deeply ethical questions. What types of bodies are assumed in gender and development debates? Who speaks for them? Whose bodies matter? How can we move beyond the current 'empowerment-lite' gender and development regime? Could you expand a bit on how the body politics framework relates to issues of space and scale? How did Doreen's work affect your thinking through of the framework?*

Wendy: Perhaps the most important observation I can make in response is the following: Our fractured global landscape is visibly and interestingly criss-crossed by connections, forged by ordinary people who do not just await their fate. In my book, I referred to a whole range of actors who are engaged in body politics and development attempting to work across divides, using policy instruments but also searching to reach out to the embodied realities of the discrimination, violence and pain as well as the pleasure, the care and the love. No one is engaged in such debates innocently. It is important to question all of our roles in a world marked by a development that has led to climate change, water and food scarcity, oil crisis, spiralling consumer costs and deepening gender inequalities. Those questions are the source of vision, networking and connection with others around the world.

My book *Body Politics in Development* locates gender and development within the contestations around the body, with the idea that the body is often the first entry point into political space for women. Gender and development's focus on the body opened up ways to speak about unjust and painful practices; however, it also normalised and silenced women's different bodily experiences. I was wanting also to push beyond heterosexist, racist, neocolonial, culturally blind understandings of violence, sexuality, care, biotechnologies and other issues which are not sufficiently nuanced around diversity of place and the resistances of women (and men and transgenders) to normalising development projects. I wanted to show, going beyond 'empowerment-lite', that people working in gender and development in situations of relative privilege can and do speak out about violent and unfair situations experienced by others in strategic ways, oriented towards finding constructive solutions, without deepening marginalisation and discrimination.

My idea was that body politics link some spaces in development and in the resistance and creativity of women's movements in ways that could move beyond dominance, and make connections. In these connections, organisations and communities find ways of dealing with deep bodily pain in a manner that attests to concern and commitment, making a difference for all involved. By conducting dialogues across the global North and global South, global and local, academe and activism, and diverse identities and cultures, body politics can move out of institutional power games, jargon and over-theorising, media hype, individual narratives and romanticism.

I was striving to set out how feminists should collectively seek to understand multiple differences and asymmetries of power, use the insights from differences, hear the anger, note the silences and keep conviction. The book's discussion of reproductive rights and health, care work, migration, sexual and gender-based violence, sexuality and techno-science shows that body politics from the community to the international level is crucial for social justice in the ways that Massey sets out.

In terms of scale, there has been a large amount of informal networking among sexual, health and reproductive rights, LGBT, health movements and HIV and AIDS activists. Through all these connections, they have inserted their concerns into major international agendas (e.g., gender-based violence into the UN conference on population), thus providing a framework to support place-based battles (domestic violence, rape in war etc.). I see the potential of transnational feminism when it moves beyond binaries, fully aware of the intersections of class, caste, race, gender and geographical/post-colonial divisions. Massey's theoretical work dissolving binaries and opening the dimension of multiplicity (what she calls the most crucial aspect of space) invaluably informs an empowered 'gender and development', making visible women and men's layered lives, roles and responsibilities (Massey, 2004: 14). Embodied realities, quality of life, rights and access to resources, as well as pleasure, cultural diversity and expression, would be factored in, not as additional but as integral to a social justice agenda.

Alice Brooke Wilson: Food Politics and Geographies of Sustenance

My work in and on the 'local food movement' in the United States led me to Doreen Massey. When I first came across her work, I immediately felt, perhaps more than anything, gratitude that someone had so clearly theorised the problems I found myself confronting in the movement, namely the risk of calling for a return to a romanticised, depoliticised 'local'. Simply eating 'what your grandmother's mother ate', as leading US food writer Michael Pollan advocates, obscures serious historical inequalities. For example, North Carolina's local agricultural system was built on slave labour, sharecropping, and the unpaid family labour of women and children. The current exploitation of undocumented farm-workers has largely maintained these dynamics. This is not a system to romanticise or recreate, although nearly all food was produced and consumed locally.

However, despite these risks of oversimplification, I maintain that a movement based around integrating ecological, social and ethical commitments with food and agriculture practices has the potential to fundamentally challenge not only the direct injustices of the industrial food system, but perhaps also the structures of modernity this food system supports. I broadly define the contemporary US food movement, traceable to the 1960s 'counterculture', as people working both in opposition to the forces of industrialised, corporate-dominated food production and those committed to building alternatives. This definition includes – but is not limited to – bourgeois efforts to preserve the aesthetic quality of so-called 'traditional foods', as embodied by Slow Food; attempts to mobilise middle- and upper-class consumers to 'vote with their forks'; initiatives in low-income urban areas to reconfigure food production as a tool for community-based wealth generation and independence from the food and pharmaceutical industries, often centred around community gardens; the so-called 'locavores', middle-class and tech-savvy white ethnics who obsess about '100-mile diets'; and urbanites and suburbanites who have gone 'back to the land' to launch micro-farm projects that rely on past privilege and rarely maximise their earning power. The movement can also be stretched to include activists for farm-worker and meat-processing plant employee rights, fair trade advocates and animal rights activists. Sometimes the different subgroups work in coalition, and at other times they disagree; the movement is broad-based, contradictory and internally conflicted.

The most visible part of this movement (or set of movements) is explicitly organised around promoting the production and consumption of 'local food'. However, applying Massey's understanding of space as the dynamic simultaneity of social interrelations at all scales, and place as a particular articulation of those relations, threatens to destabilise the core element of the local food movement – the 'local'. The discourse of local food currently connotes all the elements traditionally ascribed to place contra space: authentic, singular, fixed, with an unproblematic identity (Massey, 1994a: 5). As Massey notes, this conceptualisation of place depends on a static and abstract view of space. Instead, recognising the 'inherent dynamism of the spatial' reveals the spatiality of power, and thus how space is implicated in both history and politics (Massey, 1994a: 4). I want to emphasise that the theoretical move to reconceptualise space as social, and thus the social as spatial, requires both a reconceptualisation of place and different ways of politicising space. The political potential in this movement, and I would suggest social movements in general, is crucially made visible by Massey's argument that space is not an absolute independent dimension but instead 'constructed out of social relations' (Massey, 1994a: 3). From here follow several consequences for the local food movement.

First, this movement risks marginalisation if it misreads 'local' as merely resistant to global forces. This misreading quickly leads to an overly simplified local (good) versus global (bad) – an 'exonerated' localism based on romantic essentialism (Massey, 2004: 11). In fact, some of the 'buy local' campaigns inadvertently support inequality by supporting traditional American agrarian structures, whereby landownership and labour arrangements mirror the racism and sexism that historically define US agriculture (Allen and Wilson, 2008: 537). In *Agrarian Dreams: The Paradox of Organic Farming in California* (2004), the geographer Julie Guthman drives this point home. She demonstrates that in California, agriculture has been industrial scale and dependent on exploited migrant labour for generations. The explosive rise of organic agriculture over the past 20 years, she shows, has not so much challenged this order but sanitised it for 'conscientious consumers'. Organic farms in California rely as much on undocumented workers as their conventional peers, and pay them just as little. In California as elsewhere, buying local and even organic can mean supporting – and benefiting from – an exploitative, plantation-style agriculture regime.

A second question also emerges: what does a movement based around local food represent? A triumph of neoliberal logic that valorises reactionary individual consumption practices and unwittingly reinforces the order it seeks to challenge, or a libratory political potential indicated by the politics of space implied by this movement's active reimagination of 'local'? To what extent do different ideas about the future introduce a new dynamic between space and place? Massey's work is crucial in my approach to these questions.

Questions for Alice Brooke

What are the elements of Massey's work that help you to better describe and understand the food systems and local food networks that you have been studying? What are the elements of these networks that most resonate with or challenge her work? Second, as part of your dissertation, you have written about what you call 'geographies of sustenance', linked to alternative agricultural and food systems. Could you explain a bit more about what you mean by geographies of sustenance? What specific articulations between place and space do you envision for these geographies?

Alice Brooke: First of all, a movement that has taken an explicitly spatial strategy – 'local' – to describe its goals requires serious analysis of its conception of space and place, particularly to avoid a romanticised local, or a sense that local is good simply because it's local, what Massey calls an 'a priori politics of topographies' (Massey, 2005: 172). A fundamental problem is how this spatial strategy obscures the myriad of political issues at stake in the food system, from stagnant/declining wages to agribusiness subsidies, to the health and environmental injustices associated with agroindustrial and agrochemical production. The complexity of these issues is close to home for me for a number of reasons, particularly through my commitment to the small organic vegetable farm where I grew up in the mountains of North Carolina.

Massey provides an invaluable theoretical intervention by rethinking the relationality of the local and the global. By replacing dualistic conceptions of space and place with an understanding of the inherent dynamism of spatiality, Massey suggests a theoretical framework that can fruitfully be applied to other dichotomies in addition to the local/global and male/female binaries she explores, particularly, culture/nature. Her insight into the intrinsic relationality of the spatial – which makes 'place' an articulation of flows, 'a meeting-place' – is, to my mind, her most important contribution to imagining alternative futures and thinking about social movements; this is relevant to the various concerns around gender and the body (Wendy), nature (Dianne), culture (Arturo) and of course food (Massey, 1994a: 5). Massey explicitly takes on the question of local food in *For Space*, in a discussion of Jose Bové *et al.* and the struggle to keep hormone-laden US beef out of France. Massey suggests that by making geographical diversity itself a positive value, this 'pro-local' movement manages to avoid romanticising the local (Massey, 2005: 171).

Crucially, understanding space as relational also requires shifting our understandings of time away from fixed teleologies: 'if time is to be open then space must be open too' (Massey, 2005: 59). Recognising the multiplicity and openness of the spatial requires rethinking the temporal and challenging the hierarchy of time over space; as Massey writes, 'the multiplicities of the two can together be a condition for the openness of the future' (Massey, 2005: 89). Bringing these insights together, I would say that my research on alternatives to industrial agriculture could be considered part of the anthropology of the future, through the study of ways that ideas about the future influence and give meaning to contemporary (and past) social practices around food and agriculture. My current research focuses on the social practices of imagination that social movements enact to create alternatives to the singular notion of the future imposed by agro-biotechnology.

This brings me to the question of geographies of sustenance (in direct homage to Massey); by this I mean the spatial reinscription of agriculture and food practices in such a way that they are embedded into ecological, economic and ethical concerns. Part of why I am committed to studying food is the sheer volume of trajectories enrolled by the daily need to eat. No matter where you start in the food network, and no matter which trajectories you follow, you will be faced with manifold ecological, political and ethical options and concerns. The industrial food system certainly has its own geographies and imaginaries – not really of sustenance, but of profit. My research aims to contribute to understanding how sustainable food imaginaries – which I propose are generated at the conjuncture of social and environmental imaginaries – take shape and become embedded in natural and social practices (see Appadurai, 1996, on social imaginaries, and Peet and Watts, 1996, on environmental imaginaries). I would argue that food has the potential to become a central domain for imagining alternative futures and for embedding these imaginaries in ever-expanding geographies of sustenance in the present.

However, I see real risks involved with imagining 'local' as the opposite or the solution to the industrial food system. When defending the local becomes the primary political project, then abstract space has triumphed and with it has been eclipsed the possibility for alternative futures. Massey points a way out by focusing our political attention on 'the relations which mutually construct [both the local and the global]' (Massey, 2005: 83). In my experience, most US-based local food movements ignore the relations that constitute the local and the global, or at best accept them as a given; Massey urges us to take them as an object of dispute, interrogation and study. Taking seriously a relational understanding of space also means tackling the historical trajectories inhered in each place – something at which the US local food movement has not excelled, perhaps as part of a generalised historical amnesia.

An example from close to home: the Carrboro Farmers' Market, a place literally buzzing with life and fragrant with fresh produce and flowers on Saturday mornings. There is much to admire about the Carrboro market; in a world of faceless, generic food, this place brings Carrboro residents face to face with human-scale, ecologically minded farmers selling fresh and nutritious produce. 'Buy local' rhetoric proliferates, on T-shirts and on bumper stickers in the parking lot. Given the conviviality of the scene and the quality of the produce, celebration of the local seems appropriate. Yet the scene has very little ethnic diversity; often, some of the only non-white faces on view belong to a family of multi-generation African American farmers who sell high-quality organic produce and meat. (Most of the other farmers are first-generation white farmers who moved 'back to the land' in the 1970s and 1980s as a reaction against the post-industrial economy.) The farm-stand of the lone African American family is quite popular. I've found out that they farm in a rural African American community over 30 miles away (the market limits vendors to a 50-mile radius), where most people have long since been driven out of farming by the economic challenges of the business in the United States. In order to keep their farm alive, this family charges prices that people in their economically poor rural community cannot afford to pay. Instead, like most of the farmers at the market, they truck their produce into Carrboro, where their prices are quite attractive to middle-class white consumers. In this case, the emergence of 'local' food in Carrboro precisely mirrors the absence of local food in low-resource rural communities, places where Wal-Mart tends to be the primary food supplier. The situation is of course complex; access to the Carrboro Farmers' Market essentially preserves the last family farms in these rural areas. Interrogating the case of Carrboro Farmers' Market's lone African American vendors leads easily to a critical look at Carrboro's history – its trajectory from a working-class African American mill town to a predominately white, affluent part of Chapel Hill, with a few, largely hidden, low-income pockets. Fundamentally, I suggest a critical assessment of the construction of 'local' in Carrboro. Opening up questions of food justice, including relationality across time and space, could help lead to a more inclusive, broad-based food movement in the area.

Thus, taking relationality seriously implies a geography of politics based on our constitutive interrelatedness (Massey, 2005: 189). This lesson is one of the most enlightening for me: the real importance of the multiplicity of space is 'the coeval multiplicity of other trajectories and the necessary outwardlookingness of a spatialised subjectivity' (Massey, 2005: 59). Massey's radical thesis, that everything – especially hope for the future – depends on relational space, means that geographies of responsibility extend not just into the past, but also into the future (Massey, 2005: 195).

Arturo Escobar: Relational Ontologies and Geographies of Responsibility

I first came across Doreen's work in the mid-1990s. I must have been looking for a good anchoring point for the political. Reading 'A global sense of place' – such a counter-intuitive, even oxymoronic, title, I thought at that point, but of course she was absolutely right, in that all senses of place are inevitably inflected by the global – I realised that such a potential existed in the notion of place; it immediately appealed to me, grabbed me. I read a few more of her pieces of that period which dealt with her evolving thought on the relation between space and place, and 'the power-geometries' they both enact and exist within. The work I had been doing in the Colombian Pacific was characterised by a strong territoriality, which I found eloquently expressed in activist discourses. In this double context – my work in the Pacific, and Doreen's work – 'place' clicked. Over a decade later, it still does, and my understanding of place has since much improved, in large part thanks to Doreen's work. Round about 1997–1998, I also read, with the friends in the WPP group, a yet unpublished (and later on, well-known) piece by Arif Dirlik, 'Globalism and the politics of place' (2001); this piece was also influential in our thinking, and it drew directly on Doreen's work.

If I can make any claim to 'being a geographer', I owe it initially to Doreen. I had engaged, of course, with a lot of work written by geographers (mostly of the Marxist persuasion, including David Harvey's work), but I read those works as political economy and as focused on something a bit strange for an anthropologist, called 'space'. Even with the (timid) 'spatial turn' in anthropology that started in those years, I did not see the work I was doing in geographical terms. Doreen's combination of place, feminism, political economy, and a particular political sensibility resituated my questions within the lively debates in geography. In short, her feminist politics of place showed that, without shunning political economy, it is possible to find profound meaning and hope in the struggles carried out by peoples in places, and that this can be seen as 'a local politics with a wider reach' (Massey, 2004: 11). This was close to what we called WPP. I have since learned much from other feminist geographers, particularly J.K. Gibson-Graham (also participating in the WPP project) and Dianne Rocheleau, and from other geographers, but my initiation into the field came directly through Doreen.

Three of the most important lessons for me from Doreen's work were, first, the need to always think of place within networks of relation and forms of power that stretch beyond places; second, that places are always the sites of negotiation and continuous transformation; and third, that any relational notion of space and place ineluctably calls for a politics of responsibility towards those connections that shape our lives. I learned, in short, about the need to avoid falling into the trap of 'exonerating the local', which Wendy exemplifies well in her discussion of the disembedded body politics in development; about the acute awareness of how groups of people themselves always reconstruct places, and what this has to do with translocal forces, diverse interests and complex spatial processes; and about the ethics of connectedness that follows from a relational conception – what Doreen calls 'geographies of responsibilities' (Massey, 2004).

This notion of 'geographies of responsibility' has become more prominent for me. Once one adopts a relational notion of space and place, she argues, then all politics has to be a politics of connectivity, with the implication that one has to acknowledge one's responsibility in all of these connections; Alice Brooke exemplifies this point very nicely for the case of food and the Carrboro Farmers' Market. A second question of continued interest for me, as an anthropologist, is that of difference. I always seem to want geographers to deal more with 'the cultural' in their theories and this has been the case with Doreen as well. My most persistent interest is that of the politics of difference, and today I am trying to articulate it from the perspective of ontological difference, leaning on the notion of relational ways of being and doing. How can we talk about this difference without slighting the links of place with capital, power and the state?

Questions for Arturo

In your most recent writing, in Territories of Difference *(2008) and beyond, you have been talking about networks, assemblages and, above all, relationality. Do you find that these relatively new frameworks resonate with Doreen's work, extend it, or are in tension with it? In what ways?*

Arturo: As I mentioned, Doreen's notion of space and place is deeply relational. However, I do think there are tensions between the approaches on which I have been drawing most recently and Doreen's way of thinking about relationality. To cite an example from within geography, the debate generated by Marston, Jones and Woodward's 'Human geography without scale' (2005) suggests the possibility of doing away with scale altogether. I think Doreen would go along with this critique up to a certain point (certainly on the 'phallic verticality' of much scalar thinking), but not all the way through to the dissolution of scale. And there are other tensions, for instance, between Deleuzian/Guattarian approaches to territorialisation versus Doreen's space, which Doreen discusses in *For Space* (2005). But nevertheless, Doreen's argument about an ethical politics of connectivity is not only applicable to those other frameworks. To put it bluntly, there is much that all of us working on relationality have to learn from Doreen's conception of 'geographies of responsibility'. Allow me to explain.

Let me first highlight some aspects of Doreen's geographies of responsibility. First, a politics of responsibility is a sequitur of the fact that space, place and identities are relationally constructed. We are all implicated in connections, and we must have an awareness of this fact of such a kind that enables us to act responsibly towards those entities with which we are connected – human and not. Analysis of these 'wider geographies of construction' (Massey, 2004: 11) is central to this awareness (let me say in passing that not only academics engage in this kind of analysis; it has become central to many social movements, and activists oftentimes have a deep sense of their relationality vis-à-vis the social and natural worlds). Second, we need to be mindful that 'a real recognition of the relationality of space points to a politics of connectivity … whose relation to globalisation will vary dramatically from place to place' (Massey, 2004: 17); this calls for some sort of ethnographic grounding to that politics (in a broader sense of the term, that is, in terms of a substantial engagement with concrete places and connections). Third – and this is perhaps a point little noticed in this well-known piece – the 'geography of our social and political responsibility' that emerges from relationality also leads us to ask: 'What, in other words, of the question of the stranger *without*' (Massey, 2004: 6, emphasis in the original), of our 'throwntogetherness'? Which, of course, ineluctably links up to issues of culture, subjectivity, difference and nature. The following quote sums up for me these notions: 'The very acknowledgement of our constitutive interrelatedness implies a spatiality; and that in turn implies that the nature of that spatiality should be a crucial avenue of inquiry and political engagement' (Massey, 2005: 189).

Let me now turn to some of the emergent ways of talking about relationality – surely a timely trend. Post-constructivist perspectives on relationality are springing up in a broad variety of intellectual and political terrains – from geography, anthropology and cultural studies to biology, computer science and ecology. Some of the main categories affiliated with this diverse trend include assemblages, networks, relationality, non-dualist and relational ontologies, emergence and self-organisation, hybridity, virtuality, and the like. The trend is fuelled most directly by post-structuralism and phenomenology, and in some versions by post-Marxism, actor-network theories (ANT), complexity theory, and philosophies of immanence and of difference; in some cases they are also triggered by ethnographic research with groups that are seen as embodying relational ontologies or by social movements who construct their political strategies in terms of dispersed networks. Taken as a whole, these trends reveal a daring attempt to look at social theory in an altogether different way – what could broadly be termed 'flat alternatives' (see Escobar, 2010, for a more adequate treatment of these trends).

In geography, some of the key interventions are the debates on post- and 'non-representational theories' (e.g., Pickles, 2004; Thrift, 2007), 'hybrid geographies' (Whatmore, 2002), 'human geography without scale', already mentioned, 'rooted networks and relational webs' (e.g., Rocheleau and Roth, 2007), and the shift from dualist to relational ontologies (e.g., Castree, 2003; Braun, 2008). These trends build up a complex argument about scale, space, place, ontology, and social theory itself; 'nature', 'ecology' and 'politics' are often (not always) present in these debates, most potently in Whatmore's and Rocheleau's cases. In these works, there is a renewed attention to materiality, whether through a focus on practice, or relations, networks, embodiments, performances, or attachments between various elements of the social and the biophysical domains, and to bridging previously taken-for-granted divides (nature/culture, subject/object, self/other) into processes of productions and architectures of the real in terms of distributed socio-natural formations. Space is no longer taken as ontologically given, but as a result of relational processes. This links up with Doreen's work, although, as I hope should be clear even from this cursory review, through quite different languages.

Is Doreen's notion of connectivity the same as in some of these works? I believe only partly so. One may ask: Is the geographer's 'relationality' the same as that of the phenomenologist, the indigenous activist, the ecologist, or the assemblage theorist, to mention just some of the most prominent perspectives from which the concern with relationality is stemming today? There is a sense that Doreen's and some of these works speak of a different geometry of power and being. However, the main point for me is that, regardless of how one answers the question, Doreen's emphasis on the ethical and political implications of relational views are eminently applicable to all of the above trends. Thus when she speaks about 'the possibility of a more extended relational groundedness' (2004: 10), and of 'the challenge of our constitutive interrelatedness' (2005: 195), she illuminates for us all the kinds of politics that relationality and connectivity entail – no matter how theorised, or to what range of beings and situations these politics might refer.

Dianne Rocheleau: Rooted Networks and Geometries of Power

As a geographer trained in both physical geography and the nature/society focus, I ventured into development, Latin American studies, systems ecology, political ecology and, later, feminist political ecology and practical political ecology. I first seriously engaged with Doreen Massey's work at the suggestion of Arturo Escobar, Wendy Harcourt and Julie Graham in the process of the WPP project. Doreen's writings on place, space and relationality have served as a way to ground and expand my emerging work on networks, complex systems and relational webs. Especially crucial to my recent thinking is her combination of the geometries of power with networks and relational concepts of space and place. Three of her works – *Space, Place and Gender* (1994a), *For Space* (2005), and the article 'Geographies of responsibility' (2004) – have made it possible for me to relate the insights that I have found in complexity and network theories to the space and place domains of geography. Through reading Massey, I have gone beyond reconciling networks, power and territories to question our given notions of each, especially with respect to the problematic gendered binaries of place/ space and nature/culture.

In the past, rooted in the environment/development nexus of globalisation debates, I understood every map to be inherently a negotiated settlement. I saw each map as one reconciled image showing a set of perspectives on what-where-when in relation, not a description of a pre-existing place in space. My ideas changed once I engaged with Massey's language, clarity and insight. Her probing questions and insightful connections in the three works noted above, and her commentary on our WPP work (Massey, 2002) helped me to rethink my ideas on rooted networks. Massey has left me with the persistent and thought-provoking question about the essentialised, gendered, dichotomous views of space and place still present in much of the scholarship on social movements as well as various environmentalisms.

After our work on WPP, just when I thought I might fall again into the chasm between place-based social movements and political ecology, feminist post-structuralism and local versus global connections, her work on care and responsibility provided a lifeline out of the abyss. She makes clear that the question is not about choosing between working 'at home' or 'over there'. Rather, our challenge is to explore, explain and act upon the multiplicity of relations within and between many homes and many distant actors, as well as the structures that shape, and are shaped by them, across scales. Her presentation of this approach helped me to clarify my own position, to reconcile thinking, writing and acting in Worcester, Massachusetts, San Cristobal de las Casas in Mexico, Ukambani in

Kenya and Zambrana de Cotui in the Dominican Republic without succumbing to, or denying, the entangled and unequal relations within and between them.

Where Escobar notes that culture is still not as fully present as he would want, I find that 'nature' and ecological questions need to be more central to the conversation. Territories, while porous, also need to exist (Dirlik, 2001) in order to mediate selected access and protect the integrity of autonomous cultures, ecologies and politics, however fluid and dynamic they may be. But then that is our job: to take what Massey has done so well and to fold those insights into the mix of ideas and practices that we all construct, individually and collectively. I see my own challenge in the need to reimagine the geometries and politics of power in networks, to remake our vision of the global–local nexus that both roots and limits us. I find myself drawn to concepts of rooted networks, shot through with power, embedded in nature and culture, anchored in the fluid space-time that Massey has traced so elegantly in her various works.

In recent empirical research in the Zambrana-Chacuey (Dominican Republic), Chiapas (Mexico) and Worcester (Massachusetts, USA), I explore the relation between sovereignty and autonomy, as well as the terms of connection between elements in networks and between whole networks and territories/spaces/places. Territories may be defined by extraction, disposal, residence, refuge, commerce, creativity – and always, encounter and relation. In particular, I am compelled to investigate and theorise the geometries of power in combination with the terms of connection and the geometries and terms of rootedness, including ecological relations of power in place(s) and space(s).

As I go forward, I will certainly travel in the company of the WPP group on culture, places and globalisation, as well as Arturo's work on networks and relational ontologies and his reading of several theorists on complexity (Stuart Kaufmann, Marilyn Strathern, Enrique Leff, Humberto Maturana and Francisco Varela, and Boaventura de Sousa Santos). Likewise, I will continue to draw upon Wendy's recent writings on body politics and place, the work of the WPP group and a successor group on feminist political ecology, Alice Brooke's careful rethinking of scale and justice in food systems, and Massey's collected works on care, responsibility, place, space, gender and justice.

Questions for Dianne

The notion of 'rooted networks' is wonderful perhaps because it's so counter-intuitive, in that networks have been about flows and mobility above all else. It's doubly important because many 'network' approaches have so little place for ecology, or the biophysical dimension of the networks. Often, only those biophysical elements that articulate directly with capital or the state seem to matter! Can you say a bit more how in your work you have come to see 'geometries' in terms of 'rootedness', and the role of ecology in it? How do we infuse space and place with a greater ecological sensibility?

Dianne: The spaces and places of Zambrana-Chacuey in the Dominican Republic which both created and were created by the Campesin@ Federation of Zambrana-Chacuey, challenged me to go beyond the ideas of biodiversity and campesino/a solidarity and to look at differences within groups and between groups, in place(s) and between places. New spaces were clearly created by these relations but I couldn't articulate this critical approach to spatiality without the kind of language and theory that Massey brings to bear. Likewise, I was seeing assemblages somewhat similar to the types that Latour describes but without the language to talk about care and responsibility, about the powers of connection and the connectivities of multiple kinds of power. Further, as I worked with women's and men's groups, and various overlapping organisations in the liberation theology, land struggle and alternative development movements, separately and together, I left the comfortable territories of identities for the complex entanglements of identity, affinity and relational webs woven from assemblages in place and across places.

Field research with Federation members and groups over a period of 15 years (1992–2007) allowed us to 'see' a gendered and formerly invisible, regional agro-forest in a finely textured mosaic landscape that had been characterised by the Forest Department as a 'deforested zone'. Yet our findings showed that this invisible, extensive agro-forest supported an impressive diversity of crop and forest species as well as a complex array of livelihoods among the members of the Federation (roughly 800 members and their family members [3200] in 500 households) and the larger population (an additional 8000 people) in the region. Coffee, cocoa, their shade and companion trees, timber and a profusion of fruit and 'forest trees' were woven throughout the landscape in patches and ribbons. The meshing of the people, their farms and the agro-forest was shot through with power, from all sides, embedded in national and international structures and fuelled by the energy of self-organisation from below. Likewise, the Federation itself was riddled with conflicts and power relations based on gender and class, as well as political beliefs and affiliations. Yet, for over 30 years, solidarity trumped differences. In fact, the culture of the social movement and deeper elements of regional cultural roots enabled individuals and whole communities to organise, to act and to persist in the face of powerful state and market forces, working across recognised and respected lines of difference as well as tapping commonalities and shared values.

The images that kept seeping through the dominant templates of space and place were the relational and asymmetrical geometries of power with and power against, as well as power over, power alongside and power in spite of. These same relations in place were, and continue to be, shaped by global biophysical, economic and political processes, from climate change to multinational mining companies. The Federation (founded in the 1980s as part of a cross-scale land struggle movement) is now engaged in a struggle to define itself in a context where the former Rosario Gold Mine has reopened under new management (a US and Canadian-based multinational, Barrick) and the regional water supply is once again threatened by toxic mine waste. The regional agricultural economy is also highly challenged under the recently enacted CAFTA (Central American [and Caribbean] Free Trade Agreement). The Federation and the region are both located in histories and geographies of entangled powers, woven into particular moments of space-time. What is also clear and still does not completely fit into Massey's work are the rock, soil, water, plants, animals and human bodies, the bittersweet blood and sweat of everyday ecologies that shaped and were shaped by the Federation in the regional place they call home.

The construct of rooted networks can help to address relations of power, connectivity, place and space across both social and ecological domains in places such as Zambrana-Chacuey. It draws upon social and ecological network and complexity theories to rethink the geometries of networks in relation to space, place and territory, and to deal with horizontal and vertical power relations within and across scales. This in turn provides a way to reconcile control from above/outside with self-organisation from below, within and between.

While Massey's relational geographies of space and place consider the physical environment mainly through the mutual construction of people, place and space, ecology has focused on the relation between living beings and the transformation of assemblages and associations of various plant and animal species into communities, usually apart from human beings, except as disturbances to 'natural' communities and their habitats. The ecological geometries of relationality have tended to be hierarchical. In systems ecology, this has been pictured as a layer cake of trophic relations in energy flows from green plants to primary consumers to secondary consumers, to top carnivores and decomposers, entwined with 'nutrient' (material) cycles and recycling, all in a metabolic process somehow encased in an abstract, a-spatial, biosphere bubble. Likewise, community ecology has constructed two dominant geometries of ecological relationship: first, the hierarchical chain and simplified network relations in food chains and food webs; and second, flat earth maps of territories and biomes as containers of inventories in spatially defined units. The relationship of individual organisms and species to their environments and their relations of competition and predation have been embedded in hierarchical models often focused on scarcity, limits and antagonistic interactions. The use of network geometries has been limited to food webs, until the emergence of recent work on ecological networks (Fath and Patten, 1998). What both schools of ecology still neglect is the organisation of these elements and relationships, in space and in place, and the emergence of new elements and entirely new systems. Complexity theory has just begun to deal with the dynamism and the capacity for self-organisation and ongoing restructuring, across scales and outside of the fixed hierarchies previously imagined in systems and food chain models. What is especially promising is the possibility of reconciling the work of Massey and others on space and place with that of complexity theorists, as well as the earlier biophysical and multi-scale metabolic relationality posited by systems ecologists and the networked web of relations increasingly advanced by community ecologists.

The practical politics of food in urban and rural ecologies have also helped me to rethink the geometries of power in terms of rooted networks. As an exercise in practical political ecology, I have challenged myself and my students to link the food that gets to our plates in one shared meal to questions of social, cultural, economic and ecological justice and daily practices in far-flung places of production and processing, along with the routes and processes that link them to us. For example, we are joined to the realities of rural smallholder farmers in southern Mexico by our common dependence on their land and the resources there, as well as by their labour in the production of our food and our purchase and consumption of their produce. As consumers of highly subsidised US corn products, we are also connected to their territories as agents of displacement. The exclusion of their produce from the markets in their own country and ours forces many farmers to migrate from their own lands to work in distant places, whether on the Riviera Maya south of Cancun, the dairy farms of Vermont, or the suburban lawns of South Florida. We are entangled in networks of mutual influence without taking responsibility for what we do, or what we foster indirectly through daily practice. By following the ecological and social trail of the food on our table back to the sources, and the intervening places and processes, we can define our own networked links back to overlapping and entangled territories of production as well as to land, farmers and water sources in a very material sense.

We are all always rooted, vertically and horizontally, to land and water and sunlight and nutrients somewhere and to the people, plants and animals in the places where elements of our own food, fibre and shelter are produced, or not. Entire networks are sometimes rooted in a single point on the surface of the earth, but likewise, some networks may have many dispersed roots extending like tentacles across and around the planet. We have horizontal roots that draw sustenance between networks, yet always there is a source in a territory somewhere, fuelled by sunlight, water, fertile soil and even petroleum and a socially constructed spatial frame. Roots can feed us and they can anchor us, and they can travel between networks, but they need to go into ground somewhere.

'Rooted networks' – inspired by the Rural Federation of Zambrana Chacuey as well as Akamba farming communities in Machakos District Kenya, Zapatista communities in Chiapas Mexico, as well as Massey – are a call to rejoin networks to land, territories and ecological materiality in our thinking, since they are always already joined in practice, whether we notice and take responsibility or not. Likewise there is a parallel call to place the biosphere bubbles, the food webs and biomes of ecology in socially constructed space and place, and to acknowledge the social relations of power that frame exchanges within and across networks and between complex networks and territories.

★★★

In Greek myth, the muses delivered inspiration and knowledge. What has our Massey Muse brought us? For each of us, our interaction with Massey's work has given us new ways of seeing, new vistas from which to understand and analyse the social movements that animate our work. Massey spurs us not only to embed our analyses in space and place, but also to problematise, interrogate, think and rethink the role of space and place, those superficially arid concepts that are in reality shot through with multiple histories, power relations and shifting narrative constructs, each with their own twists and turns. In a sense, she has taught the anthropologists among the four of us to become shadow geographers, fixated on the spatial components of our work; and the geographers to become shadow anthropologists, fixating on the cultural potential of space and place, their ever-shifting status as cultural constructs. In the multidimensional spatial universe that Massey has sketched, each of us is mapping out ways forward for understanding and shaping the worlds we occupy.

Yet the muse metaphor, as precise as it is in describing what Massey has done for us, is also limited and limiting: for the Greek muse's knowledge and inspiration only flowed one way. It is our hope and ambition that our work will remain in conversation with Massey's and that those conversations will continue locating and generating new practices, memories, and even songs.

Chapter Twelve
A Physical Sense of World

Steve Hinchliffe

Introduction

I write at a time when a volcanic eruption in Iceland is conspiring with
anti-cyclonic conditions over northern Europe to produce a cloud of slow-
moving ash, periodically grounding air traffic within European air space.
The ash is potentially disastrous for jet engines as it might lead to engine
failure. But grounding flights is not an easy thing to do. People are stranded,
unable to travel. Economic recovery is allegedly jeopardised. The pressure
to reopen the sky for commerce is huge and, partly as a result, the science
of air safety and regulation is constantly questioned. For a few days, though,
the skies above Europe are devoid of vapour trails and the roar of jet engines
is eerily absent for those used to living under flight paths. The earth's
atmosphere is temporarily given respite from aircraft emissions.

This interruption to the noise and air pollution above our heads spoke of
a world (or at least a corner of it) that was partially and temporarily
suspended. It was a world that was left to confront its interdependencies
and spatialities. It was also a world that felt suddenly physical. The hi-speed,
time-space compressed, globalised world of air travel was disrupted by
particles less than 2 mm across. Or, to reverse the relative sizes, the spreading plate margin where Eurasian and North American crust is generated,
where ancient Icelandic peoples used to place their parliament as a reminder
of worldly divisions, forcefully suspended human activity. Whichever way

Spatial Politics: Essays for Doreen Massey, First Edition.
Edited by David Featherstone and Joe Painter.
© 2013 John Wiley & Sons, Ltd. Published 2013 by John Wiley & Sons, Ltd.

you tell the story, this was a world disrupted by a physical event. It wasn't a world that suddenly became physical (in the sense that before the cloud the physical didn't matter), but a world revealed to be contingent on physical matters, processes and events. It was a world, as I noted, that suddenly felt physical – a physical sense of world.

In this chapter I want to focus on this physical sense of world, and on the importance of Massey's writing to a world that is material as well as spatial. Starting with a brief discussion of Massey's reimagining of place, I will move on to consider her engagement with the more-than-social elements that make worlds and places matter. The chapter aims to foreground her attempts to write that more-than-human world and to urge geographers, human and physical, to find a language and perhaps a practice that is truly geographical, that in some ways draws the discipline together. It is her commitment to a geography that deals, as she says, with the complex systems of the human *and* physical worlds that forms the focus of this chapter.

A Relational World

Looking out at a clear blue sky, strangely devoid of vapour trails, I immediately think of Doreen Massey's famous essay on a global sense of place (Massey, 1991b). Reading that concise yet wonderfully rich essay again, a number of issues jump off the page. First of all, there is the fact of mobilities. Places are linked via movements. Those movements range from the obvious (the planes and the travelling) to the less obvious (the infrastructures that draw in a huge hinterland or the footprint that places make). Second, movements between places are also movements of those places. There are movements that bring places closer together, as well as those that make places more distant. Communications and transport can link places, but at the same time others are bypassed, flown over or not connected. Massey conjures up the now familiar scene of a high street continually transformed by its linkages, its connections and disconnections. Foods, fashions, people, technologies, stories, issues– all are testament to a globalised place that is, as a result, moving. In her essay, Kilburn High Street, in London, is pulled this way and that. Some issues draw the place closer to Ireland, others firmly place the street within the social economy of India, and so on. While the histories and geographies are complex, and anything but epochal and linear, places are conjoined and are co-dependent. These movements, interactions and disconnections span continents and oceans, and, even though this is not new, it is the densities and intensities of those relations that currently demand attention. Third, and vitally, such joinings are at best partial and at worst unequal. The inequalities are legion, but it took Doreen Massey's writing to keep our eyes on the game at a time when the talk was too focused upon a fast track to the annihilation of space by time. As other social scientists wrote rather triumphantly of time–space compression, of the speeding up of the social and the dense interconnectivity of lives, Massey emphasised the different relations people had to globalisation. Acceleration for some meant deceleration for others. Movements always left others behind, or bypassed places, or made other kinds of mobility more difficult. Likewise, as she noted, not only were there 'differences in the degree of movement and communication, but also in the degree of control and of initiation' (Massey, 1991b: 26) of these movements. For example, many American cities have a tale to tell of automobile companies forcing out the public transit alternatives as private corporations sought customers for their cars and leant on public authorities to finance the construction of freeways. This combination of differential access to and control of movement contributed to what Massey referred to as 'power-geometries'. For Massey, then, there were always power-geometries to this mobile planet. Fourth, Massey reminded us that hers was not simply a moral or political point, one of recognising inequality, but it was conceptual too. Power-geometries provoked Massey to rethink place – not as a bounded space, self-coherent and closed – but as partially mobile, contested and contestable, constituted and active in linking to and interacting with other places. In this formulation, place is imagined as:

> articulated moments in networks of *social* relations and understandings, but where a large proportion of those relations, experiences and understandings are *constructed* on a *far larger scale* than what *we happen to define* for the moment as the place itself, whether that be a street, or a region or even a continent ... this in turn allows a sense of place which is extroverted. (Massey, 1991b: 28, emphasis added)

As I've noted, Massey gave us an extroverted Kilburn High Street in north London. With colleagues at the Open University there was a rethinking of the region along just these lines (Allen *et al.*, 1998). And the power-geometries of a world city have been unpacked in a similar fashion (Massey, 2007). As I will show, this extroversion is highly provocative and useful, but I have also highlighted in the above quotation some words that we need to problematise (and that Massey went on to grapple with in later writings). For in elevating the social, its larger scales and networks, as the means through which a large proportion of reality is really made, there is a risk of downplaying other matters. The things that are generally thought to be outside the social are excluded; the stuff of the world that doesn't adhere to scalar confinements, the matters that are not clearly constructed or are themselves big players in their own construction. In other words, what else is articulated in and through places and how can this extroverted sense of place allow for a more than human world? Unpacking this imagination of place starts to point us to the kinds of vital changes and challenges that

Massey set herself as she took this important sensitivity to the contextures or multiple meetings that make and are made by a place to engage with a more than human world.

Beyond the Social

Writing places as unfinished, as events, as moments, Massey's great passion for another kind of geography starts to emerge. Responding to one of her many broadcasts which had focused on a (social) global sense of place, an old school friend articulated a problem: 'That's all right when you talk about human activity and human relations. I can understand and relate to it then: the interconnectivity, the essential transcience ... but I live in Snowdonia and my sense of place is bound up with the mountains' (Massey, 2005: 131). It touched a nerve. Massey herself, passionate about cities, infectious over her will to spatialise politics, was and is also in some way 'bound' to the mountains (a frequent visitor to Cumbria in the North-West of England), in love with the seemingly regular movements of migrating birds (watching keenly for the first swifts of the year to arrive over the skies of north London) and instrumental in engaging in conversation with physical geographers over the 'nature' of landscape and geography. So what kind of 'bind' did the non-human world present for her global sense of place? For someone who had done much to unhinge place and identity from any sense of fixity or inherent nature, the last thing she would do would be to allow that other kind of nature – the mountains, rivers and trees – to fix or act as foundations for her sense of place. So I want to review her unseating of foundational natures by engaging with her historical approach to landscape. Second, I consider her understanding of space-time, which proceeds hand in hand with her rendering of science and landscape as historical. I will conclude the chapter by noting how such arguments have been used by Massey to seek to cross the divide between social and physical sciences, and between human and physical geography.

Non-foundational Natures

Beyond understanding places as 'articulated moments in networks of *social relations* and understandings' (Massey, 1991b: 28, emphasis added), there are the articulated moments with others that are not easily configured within social relations. But how to think places that are at first blush less global than Kilburn High Street, seemingly not configured by the 'now' or recent histories of wars, trade, migration, telecommunication and so on? How to think about places that seem bound to the mountains? Massey turned to her beloved Lake District, in England, that supposedly timeless of places which is often held up in tourist brochures as something of an antidote to cities like London (and of course achieved its apogee within the Romantic backlash to industrial urbanisation from the eighteenth century onwards). Noting that the Lake District was anything but timeless was hardly new. There were, and are, plenty of readings of landscape as socially made through cultural and economic histories. They allow us to see the natural landscape as deforested, farmed and owned, and can highlight the patronage of the Lake District artefact by wealthy traders and artists. But alongside this social construction of landscape, Massey purposefully engaged with that outside of most humanities and social science landscape writings, the rocks themselves. Massey refused to confine herself to the times and spaces of the conventionally historical, refused the implicit local- ism of accounts of dwelling and found the piety of a phenomenological approach to landscape insufficiently open to the otherness of that outside (Massey, 2006). She set herself a less comfortable task of unsettling the very substrate of landscape and place, and used geology to do so.

In her description of Skiddaw in the Lake District she notes that alongside the country houses, the farms, the ancient mines and so on there is the 'massive block of mountain, over 3000 feet high … ; not pretty, but impressive; immovable, timeless' (Massey, 2005: 131). The rock, she goes on, had presided over the histories of the place. Despite her first take on its permanence, on its role as a substrate for human history, turning to geology allows us to see, in a way that will be familiar to students of geomorphology, the rock's own history. The latter includes having been scraped by glaciers, laid down in ancient seas and baked by volcanic activity. Such a history may involve almost unimaginable swathes of time, but it also speaks of change and history. 'One thing it might evoke is the antiquity of things. But another is almost the converse: that today's "Skiddaw" is quite new' (Massey, 2005: 133). Moreover, newness is not the only issue. The rocks are what Massey calls 'immigrant rocks' – laid down elsewhere (in what we now call the southern hemisphere) subject to the forces of plate tectonics and always 'on the move'. '[T]he rocks of Skiddaw are … just passing through here, like my sister and me only rather more slowly, and changing all the while' (Massey, 2005: 137). In this telling, Skiddaw (the place) is as physical as it is human, but its physicality is neither fixed nor timeless. As climates shift and landforms move, places are events, constellations of more than social rela- tions. As Massey states it baldly: 'Places as heterogeneous associations' (Massey, 2005: 137).

This heterogeneity of place implies difference; it implies physical and social mixture. It also signals a complexity. To be clear, this is not place as a coherent result of separate and discernible processes, or a mixture of elements that somehow add up. Rather we have an always-unfinished con- junction of trajectories, a heady 'throwntogetherness' where coherence or stability is something that is likely to be a temporary and rare achievement. So place is not simply extroverted, it is also always in tension, never settled. In 'sharp contrast to the view of place as settled and pre-given, with a coherence only to be disturbed by "external" forces, places ... necessitate invention; they pose a challenge' (Massey, 2005: 141). Place depends for its identity, its character and uniqueness on the constellation of trajectories, its throwntogetherness, as Massey puts it, and that identity is always in process and unfixed. Never simply a surface, landscape too is a conjunction, an event, a happening, a moment 'that will again be dispersed' (Massey, 2006: 46).

Immigrant rocks almost enable Massey to throw off the last anchor of place to a foundation story. Nature is less of a bind than it may at first seem; or better, the binds are not ones of fixity and truth, but matters to work out. As we will see, there is also a need to confront and engage with another foundationalism, the nature of our knowing of nature, or our reliance on science, as she calls it (Massey, 2005). For the moment, though, it is worth emphasising that there is a political project here, to undermine those who would use foundational stories to see a place as belonging to one social group at the exclusion of others. And no amount of appeal to the hills, to timeless landscapes and so on should be allowed to stand as justification for an exclusionary politics. In addition, a physical sense of place poses the challenge not only of how to respond to the myriad of social trajectories that make a place but how to live and live well with these multiple human *and* non-human trajectories. And it invites us to consider the space-times that make a place, a landscape, an entity or identity, an always heterogeneous and politically open, agonistic matter. It is to these that I now turn.

Space-time

The conceptual matter here is not just to pose the question of place itself but to link this concern to Massey's project of challenging understandings of space as they have been developed within the social sciences, within Western philosophy and often unquestioningly adopted within geography. It is of course Massey's major contribution to have us continuously question how and why space comes to denote the static, the settled, the frozen, the representational while time denotes the mobile, the living and the real. It is her insistence on space-times, or the inability to separate space from time and the need to conceptualise spaces and times as matched pairs, that can offer so much to our dealings with a more than human world. To wit, space for Massey is open and dynamic, in flux, as opposed to a fixed territory, or a slice through time. The latter is, as she has repeatedly argued, illogical, impossible even if we were to consider how to move from one slice to the next (Massey, 1997b). If space is closed, defined by the absence of time, then it becomes an unalterable terrain. There can be no novelty, nothing new and no movement. To imagine such a space is to imagine nothing but a dead space. It is far removed from the immigrant rocks, the vibrant matters and indeterminate life that Massey sees around her. Moreover, if space is impossible to imagine as static and closed, then we need to match this to an understanding of time that is equally open, historical and irreversible. Indeed, it is this conjugation of spatial multiplicity and an historical temporality that allows Massey not only to intervene in cultural politics, in globalising economies and politics, but to offer resources to those of us who are also interested in the evolution and politics of landscapes, atmospheres, ecosystems and so on. In this respect, I want to briefly raise four issues: historical science, multiplicity, place dependency and emergence. The first three of these are what I take to be useful means of conceptualising space-time. The fourth is a word I want to trouble, but do so in a way that is more than sympathetic with Massey's understanding.

With an *historical science* there is a belief that the processes involved in the formation of an entity (be that a place, a glacier, a river) are themselves historical (in formation too and in that sense relational). That is, rather than accounting for the form or presence of an entity on the basis of a timeless, mechanical process, we abandon any latent physics envy, or the belief in timeless and universal, often reductionist, explanations (see Massey, 1999b), and trouble the divide between dynamic process and historical form. Two things are going on here. First, processes are themselves historical. Second, the knowledge we have at our disposal concerning those processes is itself contingent and liable to change. In other words, as Massey succinctly puts it, the 'harder sciences', those that are often turned to in order to provide the timeless explanation or process, *'are themselves historical'* (Massey, 1999b: 270, original emphasis). To be clear, while this willingness to question and put knowledge at risk is vital in an open democratic society (Massey, 2006: 46), this is not the same as advocating a form of epistemological relativism. Massey is, as I read it, putting forward a form of knowing and practice that is non-foundational but is nevertheless worldly. This is not a nominalist science of 'anything goes' but a statement on the contingency and partiality of knowledge – a knowledge that proceeds and substantively progresses through engagements with the more-than-human world, which is itself dynamic and open to change. As a result, just as the cherry-picking of natural or physical science for convenient truths is not an option, nor is evacuating the field of debate of any 'grounds' for normative judgement (Massey, 1999b). Non-foundational approaches to nature and knowledge necessitate a serious engagement with the world *and* a commitment to the possibility of questioning and contest. That serious engagement is anything but 'grounded' in the sense of fixed or stable, but it is 'grounded' in the sense of dealing with the complex multiplicities involved in the making of realities.

As Massey is at pains to point out, this historical understanding of processes, knowledges and forms, all of which are subject to change, is not equivalent to a statement that time alone is sufficiently responsible for a Bergsonian 'continuous emergence of novelty' (Massey, 1999b: 272). For early twentieth-century philosophers like Bergson who were interested in the Darwinian world of creativity and novelty, an understanding of the role of difference and duration in the making of the new became a matter of not only philosophical interest. Conservative histories and politics were rife, based as they were on inheritance and spatial purity. It is here that Massey's engagement with Bergson's work becomes highly productive. For Massey, time itself cannot explain novelty, and change cannot be a matter of internal processes alone, or 'a mere rearrangement of what already is' (Massey, 1999b: 272). In short, for there to be change and for there to be time, there is a need for mixture, for heterogeneity and for *multiplicity*. And for all of these there needs to be space. In other words, time requires space, novelty requires multiplicity. Following Bergson, Massey notes that 'in order to retain an openness of the future, temporality/time has to be conceived ... as the product of interaction, or interrelations' (Massey, 1999b: 274). But, she adds, while temporality is created through interrelations, it does not mean that it is the cause of interrelations. For there to be temporality there needs to be multiplicity, and for multiplicity there must be space. She rewrites Bergson in characteristically accessible prose: 'for there to be difference, for there to be time ... at least a few things must be given at once' (Massey, 1999b: 274).

This multiplicity, or a few things being given at once, conjures up a sense of the vibrancy of matter, the unfixed natures of nature, the openness of space-time. Space is not, it should be emphasised, a stage or territory upon which these matters are arranged but it too is being made and remade through this activity. The point here for my purposes is that space-time matters in terms of how things turn out, how change is brought about and how things evolve. In other words, just as we need to understand place as an extroverted process, its identity and character a result of its meetings and throwntogetherness, so we need to understand other entities and identities as dependent, at least in part, on the quantity, quality and natures of their relations. To put it another way, a thing, any thing, has a history – this is something Latour and others have done well to remind us (Latour, 2004) – but it also has geographies, a complex non-coherent set of relations that help to make it what it is (Hinchliffe, 2007). Nothing is self-contained. A river is an event, an ongoing achievement of multiple trajectories, multiple processes. Likewise, a woodland is not self-contained but replete with crosscutting processes, trajectories and processes. It is not, in another language, a closed system. The arguments have effects – the tendency to close off boundaries around natural areas, a so-called fortress nature which informs anything from woodland conservation in Manchester to Project Tiger in India, threatens to underplay the connections, the co-dependencies and openings that make that 'thing' possible in the first place. Extroverted places and natures mean that any valuing of their current states requires a geographically prolific set of engagements (even where these are hidden from view). Fortresses will not hold for long. Freeze-framing places or natures amounts to little more than social and physical death. Meanwhile, the non-coherence of those places and natures make their management a complex, more than technical matter. It takes in many places and requires an approach to nature that is spatially prolific and socially and materially heterogeneous. From the complex movements of animals and their tissues (Whatmore, 2002) to the intricate weavings of environmental politics (Featherstone, 2008), the power-geometries and material relations of place and nature have informed and reinvigorated environmental geographies and politics.

As Massey notes, we talk readily about path dependency to convey how it is that one set of changes, one social or technological choice, tends to shape the future and so limit further choice. In addition, we need a sense of *place dependency* to first of all understand the importance of space in any account of change, but also perhaps to sensitise us to the vital role that our multiple relations, trajectories, materials and other cohabitants play in making space-time and in making worlds possible. In other words, where we are is as important as when we are. Interestingly, that 'where' has now of course changed. Gone are the topographical coordinates of location to be replaced by what might better be described in topological terms. Distances and proximities are in the making and space is anything but coherent (for related work that takes this project on, see Allen, 2009).

This brings me to my final point in this section, *emergence*. In Massey's engagement with physical and human geography (Massey, 1999b) there is a latent recourse to 'emergence'. It is used, in a Bergsonian sense, to understand novelty. For Bergson, time is the continuous emergence of novelty (Massey, 1999b: 272). 'Without emergence, urges Bergson (and others), there is no time' (Massey, 1999b: 273). Yet, Massey's spatial imagination and her insistence on the matching of a sense of time with an equally open spatiality, helps us to deal with emergence in ways that avoid the kind of magical ontology that infects too many accounts of complexity. For Massey, as we have noted, novelty can only emerge if there is already interaction and therefore multiplicity and space. So, she asks, why is there this ceaseless emergence? Her answer is, of course, not that things change in and of themselves (an immanent unfolding, which would involve no novelty in that matters would be closed) but that there are myriad interactions for which space is a prerequisite and by which space-time is produced. What stays the same in this supplement to Bergson is the term 'emergence'. But in many ways we have a rather different notion of emergence now than we did if we followed the more Bergsonian route to understanding novelty. Use of the same term masks an important qualification that in turn has ontological and epistemological consequences. For emergence no longer has that predominantly temporal ring to it, whereby stuff somehow happens, appears, forms. There is, to my senses, an opening here on to a spatial and ontological politics.

Useful at this juncture is Isabelle Stengers' distinction between emergence and complexity. Emergence is, of course, a term that inhabits a good deal of complexity writing. Yet, as Stengers suggests, the term is often employed to denote 'the appearance of the unanalyzable totality of a new entity that renders irrelevant the intelligibility of that which produced it' (Stengers, 1997: 12). In other words, 'emergence' is often used to refer to ordering processes internal to a system or set-up that somehow produce novelty, but the conditions for that novelty are beyond the analytical apparatus used to understand the setting. It is close in this respect to ideas of holism. As Stengers suggests, emergence sets down prohibitions (you can't understand this by looking at that), that can give emergent matters a set of essential properties that are both time- and space-less. For Stengers, and I would suggest for Massey too, complexity, on the other hand, sets out problems: it is conceptual, marking the entwined worlds of the becoming known and the becoming knowledgeable. Massey offers a scheme for comprehending, or learning to engage with, the dynamism, vibrancy and openness of human and physical worlds. Her insistence that space-time is the product of interactions starts to give us partial openings onto the complexity of worlds, not in any determinate sense (for space-time is creative and indeterminate), nor in the sense of non-caused causes (or unanalysable wholes), but in the sense of an understanding or sensitivity to worlds where space matters and where we can hope to partially understand novelty, change and so on as the results of interactions and relations. So, like Stengers, Massey's use of terms like 'complex' (and, I would argue, 'emergence') is not a response to a lack of knowledge or understanding – a philosophical shrug. Rather it is premised on an engagement with change, one that highlights the importance of space, place and difference in the production of further difference. This is the ontological and spatial politics I referred to, which arises, it seems to me, from the agonistic negotiation of people and things. As Massey has noted, this politics

> extends the normal scope of ... political science debate, which so often not only restricts its attention to the humanly social but also implicitly or explicitly depends upon a nonhuman background that is harmoniously in balance ... That space of agonistic negotiation that is the political should be recognised as including negotiation also with that realm that goes by the name of nature. It will, moreover, be a negotiation that includes within it the very conceptu-alisation of that 'nature' itself. (Massey, 2006: 45)

Conclusions: Beyond the Divide

Massey would be the first to admit that a good deal of her early work sat firmly on one side of a divide within academia and within academic geography that separated the world into human and physical compartments. Cities, economies, labour – these were matters for the human geographers – while rivers, glaciers and landforms were matters for physical colleagues. And from the 1980s onwards, in Anglo-American and European departments, the two camps attended different conferences, had separate research publications, and had little in common. For her, though, such divisions undermined the possibility for a discipline whose practitioners shared a 'common ground: that both physical and human geography – at least in large measure – are complex sciences about complex systems' (Massey, 1999b: 266). From the late 1990s onwards she endeavoured to bridge this divide, co-organising with physical colleagues publications, conference sessions and conversations which all aimed to find ways and means of talking to one another. There are, of course, sound practical reasons for doing so in a world where physicality needs more and more attention. From grounded aeroplanes to climate change, from hubristic risk assessments of deep sea drilling in the Mexican Gulf to the catastrophic events that befall the oil-polluted coastlines off Nigeria and China, from cataclysmic floods to pandemic diseases, there is plenty of evidence of the need to combine the best of human and physical sciences. But more than resurrecting resource geography, or even arguing again for multidisciplinarity when confronting a social yet thing-filled world, Massey attempted to cross the divide by attempting to open up what 'disciplinarity' involves and what it might mean to talk and act geographically. Her answer was, as I hope to have partially conveyed, conceptual and political. Indeed, if Geography is to continue to contribute to the complex world, then Massey's ability to articulate its global as well as physical sense of place goes a fair way in the continuous evolution of that discipline's identity. It takes us perhaps from a Geography that might have been too readily described as a conservative science of surfaces to a radical science of multiple trajectories, where space is intertwined with time, and where Geography, human and physical, affords us a properly dynamic and cross-sectoral approach to complex bodies and open systems. To continue the conversation Massey initiated would be testament to her provocations and contributions to geographical inquiry.

Part Four
Political Trajectories

Chapter Thirteen
Working with Doreen Downunder: Antipodean Trajectories

Sophie Bond and Sara Kindon

Introduction

Two elements strike us as central to Doreen Massey's work – openness and struggle. For Massey, an open historicity recognises the multiple criss-crossing trajectories and constructed nature of place (Massey, 2005, 2007). Central to the anti-essentialism in this understanding of openness is her reworking of space as a sphere of multiplicity, structured by geometries of power that fix certain trajectories into our geographic imaginaries that shape meanings, exclusions and inclusions. Certain trajectories are then considered to count while others are marginalised, thereby closing down a politics of alternative possibilities. Yet, struggle against such dominance is ever present, and is at the heart of Massey's academic-political work. In this chapter we draw these two elements together, to suggest that there is much to learn from Massey's academic work and praxis, and to demonstrate the potential of Massey's ideas for reimagining the politics of race relations in a post-colonial setting like Aotearoa New Zealand.

The following section highlights aspects of Massey's thinking for this chapter. We then use these to rethink the post-colonial setting of Aotearoa New Zealand in three ways. First, we explore the trajectories of space-time and the power-geometries that shape them to demonstrate the contested, complex and messy nature of race relations in Aotearoa New Zealand. Second, we explore the nature of claims making that has arisen in the

Spatial Politics: Essays for Doreen Massey, First Edition.
Edited by David Featherstone and Joe Painter.

post-political conjuncture, demonstrating both openness and struggle in a progressive politics on one hand and closure and linearity on the other. Finally, we turn to the significance of struggle in Massey's work. For it is in her conceptualisation of ourselves as the struggle that we may contribute to a more progressive or radical politics of race relations in this country.

Myth, Politics and Relational Space-Time

Central to Massey's thinking is a spatial awareness that embraces human life as 'a simultaneity of stories-so-far' (Massey, 2005: 9). Embracing the 'multiple present' recognises that the past is never singular but rather is constituted and plural, and is always already in the present via multiple criss-crossing trajectories that coalesce in a particular place and time. As such, space is relational, constructed from interrelations and interactions and therefore always in process and never closed (Massey, 2005). Consequently, space is always political and provides opportunities for a radical politics, in which space and the specificity of place are 'creative crucibles for the democratic sphere' (Massey, 2005: 153).

Yet, as Massey notes in much of her work, the hegemonic trajectories of capital letter notions of Progress, Capitalism, Globalisation, Development and Modernisation constitute linear end points for which societies should aim (Massey, 1999b, 2005, 2007). In particular, she argues that because of the hegemony of Neoliberalism and Globalisation they are constituted as the only possible narrative and all places become differently located within these frames. This in turn, 'obliterates the multiplicities, the contemporaneous heterogeneities of space … reduc[ing] simultaneous co-existence to a place in the historic queue' (2005: 5). Space becomes homogenised, and time becomes linear and singular. People, societies and places that are perceived to be 'closer' to the desired goal are positioned higher in the global hierarchy towards being 'developed' rather than 'developing'. In Massey's terms, 'the spatial is annihilated through the re-organisation of its multiplicity into a singular temporality' (Massey, 2001: 259) – a temporality that closes down possibilities and alternative ways of being, perpetuating existing marginalisations and constituting new ones.

Thus for Massey, thinking temporality aspatially, and therefore singularly, does three interrelated things that we wish to pick up on here. First, and as noted above, it denies the ever-present and never-fixed multiplicity of the social sphere, creating a singular trajectory of progress whereby geometries of power shape relations such that any person, place, system or culture not on that trajectory is marginalised. Second, it homogenises the diversity and heterogeneity of place and the constitutive entanglements subjectivities have with material geographic loci. Third, it closes down the possibility for contestation, conflict and alternative voices to be heard. Or, in other terms, it denies the value and opportunities offered by a vibrant contestatory politics that is properly democratic.

In line with the politicisation of space, Massey (2005) argues for a progressive politics of place – one that recognises the constitutive nature of social relations; acknowledges the implications of those relations in taking responsibility for their constitutive effects; and one in which the specificity and diversity of place as the locus at which a constellation of trajectories merge is recognised. Such a framing of space/place and politics takes openness and struggle as its very conditions of possibility, and lies at the very heart of our coexistence.

The synergies with other post-structural thinkers' anti-essentialisms are clear. As such, we draw on the work of Chantal Mouffe and Jean-Luc Nancy, both of whom have influenced Massey's conceptualisations of space and politics (see Massey, 1995a, 2005). We suggest that these thinkers offer further dimensions to Massey's retheorisation of space/place. Concomitantly, Massey's work enhances the perspectives they offer.

Mouffe and Massey have published discussions of the synergies between their work (see Massey, 1995a; Mouffe, 1995; and this volume). Massey readily acknowledges that her understanding of space and a progressive politics of place will 'undoubtedly lead to endless debate and disagreement' but, like Mouffe, she notes that such contestation and dissent 'are precisely the stuff of politics and democracy' and reflect the centrality and importance of struggle for envisioning and realising more just alternatives (Massey, 2005: 103). While Massey's relationality encompasses the non-human world, Mouffe's work focuses much more on the coexistence of humans, and is underpinned by a relational ontology in which dissent and antagonism are inherent and provide the very opportunities for change (Mouffe, 2000, 2005). She argues for an agonistic pluralism, where contestation and dissent lie in the relationship between adversaries (rather than enemies) 'whose ideas we combat, but whose right to defend those ideas we do not put into question' (Mouffe, 2000: 102). What Mouffe offers Massey is greater specificity in the nature of that debate and dissent between subjectivities. That is, Mouffe's radical democracy offers a normative dimension to relationships between subjectivities, in which agonistic contestation is rested on shared ethico-political commitments to equality and liberty even though the nature of those principles remains open to contestation (see Mouffe, 2000, 2005). Conversely, what Massey offers Mouffe in this framing is a spatialisation of democratic relations (see Massey, 1995a), where relationships with non-human identities (e.g., material places) and their constitutive role in shaping subjectivities becomes central to relational spatial thinking.

Like Massey, Nancy's relational ontology can be read so as to include the non-human world (see Bingham, 2006). At its heart is an ethical dimension in being to alterity relationships that has the potential to provide for the kind of responsibility to the other that is inherent in Massey's progressive politics of place (Massey, 1991b, 2005, 2007). In addition, he provides a way of exploring the myths (as both foundational and fictional construc- tions that organise social life) that shape the geometries of power that come into play in space/place.

For Nancy, subjectivities are always seeking to fulfil the absence of a fixed identity or essence by constructing commonality (or community in Nancy's terms) as an essentialised totality. These constructions are myth because the 'essence' that unifies the signification is an impossible fiction, and yet myths also found identities. He argues that these myths are neither 'good' nor 'bad'. Rather, they are necessary to structure our being-together to cover over the uncertainty and chaos that is always already present as a result of the contingency and openness of the social. Examples of such myths are commonplace: the nation state, commonly conceived place-based com- munities, ethnic groups, as well as discourses such as neoliberalism and colonialism that constitute imaginaries that are both fictional and founda- tional. Nancy (1991: 48) writes that 'myth says what it says, and in this way organises and distributes the world of humanity with its speech'.

Nevertheless, as pervasive as myths are, they are interrupted when commonality or community is disavowed or made inoperative in moments when alterity and relationality are recognised and embraced. Nancy argues that we are each singular beings marked by difference, and yet we only know our difference in relation to others – being is always being-with others as singular-plural (Nancy, 2000). Like Massey's understanding of identity, Nancy understands that we are one but we are also simultaneously multiple because our being is constituted always in relation to others – we are both singular and plural. Thus we are always already in community – not as a social construct or myth based around an essence, but ontologically such that essentialisms as myth are unworked and thereby contested. The disavowal of myth arises in moments when difference (which marks the limits of being) is exposed to others and shared – that is, when community as ontologically given is exposed. Nancy refers to this exposure as 'com- pearance' – appearing-with or 'the sharing of simultaneous space-time' (Nancy, 2000: 65). Moreover, this sharing of simultaneous space-time is a moment of freedom – a properly democratic moment in which there is a mutual sharing and an underlying reciprocity through which difference is ontologically given and understood (Nancy, 1993). Compearance is the ultimate democratic moment – when multiple trajectories coalesce in space/ place and are shared as a 'simultaneity of stories-so-far' (Massey, 2005: 9).

Immediately, we see synergies with both Mouffe and Massey's work in highlighting multiplicity and the co-constitutiveness of space and time and that these in turn both shape and are shaped by identity and place. Here, the uncertainty, chaos, contingency and radical openness of the social is exposed and full of possibilities for change (Nancy, 1991; Massey, 2005;

Mouffe, 2005). For Nancy, these moments of freedom allow for an openness to alterity and an openness to myth – thereby providing spaces within which to contest myth's dominance and its marginalising effects, and from which to assert difference (Bond, 2011). For Mouffe (2005, this volume), this openness is the very stuff of a radical democratic politics underpinned by reciprocity, equality and justice. Massey brings to their ideas the crucial spatial dimension so that we can consider how social/political relations may embrace multiplicity, historicities and the geometries of power that shape them (Massey, 2005; see also Slater's 'spatialised democracy', this volume).

Our reading of Massey along with Nancy and Mouffe facilitates an exploration of the nature of the myths that come into play in the everyday politics of and in place and the geometries of power that shape them. It provides a lens with which to understand the myths that underlie contestation and debate facilitating an understanding of their democratic or progressive character. This provides a means to engage with the way in which space is constitutive of identities (and vice versa) in order to explore the conditions under which a progressive politics arises or might emerge. In the following, we bring some of these ideas to life through an account of Aotearoa New Zealand's race relations. This demonstrates the benefits of rethinking these politics with Massey, Mouffe and Nancy for what alternative Antipodean trajectories they might enable.

Simultaneity and Linearity in Aotearoa New Zealand

Te Tiriti o Waitangi (the Treaty of Waitangi) is perhaps the most significant document in figuring Aotearoa New Zealand's race relations. It is the basis of an institutionalisation of biculturalism between indigenous Māori and Pākehā (New Zealanders of European descent), which in turn has implications for how multicultural politics are played out. Moreover, its mere presence has led to an assertion by some that Aotearoa New Zealand has 'the best race relations in the world' (Sinclair, 1971, cited in Byrnes, 2006). Te Tiriti effectively ceded sovereignty to the British Crown while guaranteeing certain rights to indigenous Māori (henceforth the Treaty and te Tiriti are used interchangeably).

Considered by some as a founding document and by others as a quaint relic (Byrnes, 2006), te Tiriti o Waitangi was signed on 6 February 1840 at Waitangi, in the far north of Aotearoa, by representatives of the Crown and iwi or Māori tribes. Two versions of the Treaty – one in English and one in the Māori language (te reo Māori) – were taken to iwi around the country and some 500 signatures were gathered (Orange, 2004). Not all chiefs signed, and of those who did the majority signed the Māori version. This is significant because the Māori version was not a direct translation of the English.[1] Thus what was exchanged and guaranteed in the English version differed from that of the Māori. Although specific implications of the Treaty would inevitably have been different for different groups, chiefly signatories broadly understood that their connection with the land as Tangata Whenua (people of the land) would not be interrupted. For them, te Tiriti represented a power-sharing arrangement as equals, so that two peoples with different cultural practices could cohabit in Aotearoa (Fleras and Spoonley, 1999). These distinctions mark out just some of the simultaneous stories that are still contested and therefore in progress at the present time – as will be seen below.

In Nancian terms, the signing of the Treaty can be read as instituting or figuring the myth of colonialism and imperial dominance in Aotearoa New Zealand. This reading does not deny the very real consequences of colonial rule, but rather draws attention to the idea that the colonial trajectory was not a given, inevitable outcome, but a series of particular decisions, in which colonialism was imagined, articulated and realised, foreclosing other possibilities.

The Treaty founded the fiction of colonial rule, creating subjectivities, communities and institutions defined around an essential totality – colonial New Zealand – set on a linear path of Progress in the image of its mother country. When complemented with Massey's concept of power-geometries, we can see how colonial and imperialist spaces provided the means through which the myth became self-perpetuating – identities became enrolled in the myth's 'truth', and complicit in the ongoing representation and actualisation of the myth's imaginary, shaping places and identities. Thus myth making has worked to maintain the geometries of power that support it, merging alterity (Secomb, 2003) into a linear homogeneous trajectory, and thereby providing for 'political institutions to gain their purpose and their legitimacy' (James, 2005: 340).

Yet this myth has been interrupted again and again by individuals and groups, marking out the struggle to assert and expose the multiplicity and alterity that was always already present but subsumed in colonial power-geometries. The land wars of the 1800s are but one example – bloody battles over land and identity and perceived breaches of te Tiriti (Orange, 2004). These conflicts and the implications of the fundamental differences in the two versions of the Treaty also resulted in it being declared a 'nullity' in 1877 by a Chief Justice (Kāwharu, 1989: x), facilitating imperial power to conveniently ignore the guarantees it provided Māori, simultaneously quashing Māori struggle to protect their land, culture and traditions.

This institutional 'forgetting' of the Treaty for a while cemented the linearity of imperialism. It also denied the existence of a relational Māori ontology within which the framing of Māori space-time indicates that Māori face, and simultaneously stand, within the present and the past (Walker, 1982), thereby 'living deeply in time, so that everything in the physical world provokes remembering' (McKay and Walmsley, 2003: 90).[2]

In this relational ontology, the act of remembering pertains to the centrality of relationships within Te Ao Māori (the Māori world), relationships with people (past and present), with non-human elements such as the Treaty, and with place. So for many Māori (and some Pākehā), te Tiriti lives on in the present. Indeed, lobbying and activism around it has been ongoing over the last 170 years, resulting in the enactment of the Treaty of Waitangi Act 1975 (see Orange, 2004). The institutionalisation of biculturalism in Aotearoa New Zealand that followed recognises that the Crown is in partnership with Māori, and must adhere to the principles of the Treaty of Waitangi. It also opened up a claims process, which we explore in more depth below.

While the institutionalisation of this bicultural politics marks out two main trajectories (or myths) – a dominant colonial history and an embattled indigenous history – these should not be reduced to singular opposing narratives, for the terrain is highly contested.

Since the 1970s, successive governments in Aotearoa have made a choice to pursue biculturalism as an official state policy with the effect that other minority groups 'are frozen out of the debate on the identity and future of the country and disenfranchised with respect to the politics of multiculturalism' (Thakur, 1995: 271–272). Yet, institutionalised biculturalism is frequently positioned in a linear temporal relationship as a necessary precursor to official multiculturalism, even as everyday multiculturalism destabilises such a conceptualisation (Huijser, 2005).

In addition, others argue that official biculturalism has not yet gone far enough in terms of equitable power sharing between Māori and Pākehā, and yet others contend that anything beyond 'official' biculturalism is impossible because of the diversity of cultures represented within the Pākehā majority (English, Irish, Scottish, French, Dutch, Italian and North American, to name a few).

The whole arena within which the Treaty claims process has arisen is inherently unstable, contested and messy (Hill, 2010). This complexity is noted because it demonstrates 'a simultaneity of stories-so-far' (Massey, 2005: 9) in Aotearoa New Zealand and the messiness of the power-geometries that tend to privilege in a hierarchy, Pākehā, then indigenous Māori, then various other minority groups. Here, we privilege attention to institutionalised biculturalism and its implications in claims made under the Treaty of Waitangi Act 1975.

Claims Making – Open and Closed spaces

Legislation in 1975 (amended in 1985) provided a means for Māori to make claims for redress from the Crown where they could establish the Treaty had been breached. This legislation and settlement process marked the beginning of an institutionalised biculturalism in Aotearoa New Zealand, through which Māori have had numerous successes over the last 50 years, both politically and in gaining compensation for past wrongs. It is often this process which fuels New Zealand's image of positive race relations.

Under the scrutiny of Massey's thinking, however, the Treaty of Waitangi Act 1975 is notable for prescribing two processes for making a claim that are contradictory. On one hand, the Waitangi Tribunal, as the institution set up to effect the Act, investigates claims and makes recommendations to the government on settlement. It is characterised by a radical agonistic and progressive politics of place. On the other hand, the formal settlement process between Māori tribal groupings (iwi) and the government emphasises 'settling' past wrongs once and for all. The settlement process lies firmly within the institutional apparatus that is embedded in aspatial linear trajectories of colonialism and neoliberalism. It follows a strict process of negotiation, followed by agreement in principle, ratification by both parties, and then formal legislation (see www.ots.govt.nz/).

The openness of the Waitangi Tribunal

The Waitangi Tribunal's task is to interpret the principles of the Treaty, to hear claims by iwi that the Crown breached these principles, and to make recommendations to the government in regard to redress. The Tribunal has interrupted the linearity of the dominant colonial trajectory by adapting its processes to facilitate meeting claimants on their marae (the centre and home of traditional and often contemporary Māori life), observing their kawa (marae protocol). It is at the marae that people, place and histories are entwined, where spirituality is expressed and people's dignity is heightened.

In addition, the Tribunal has spoken 'boldly on matters within its jurisdiction' (Sorrenson, 1989: 165), suggesting in the case of one claim that the guarantee in Article II of tino rangatiratanga (chieftainship) over fisheries as part of an iwi's taonga (treasures) was unambiguous. Rather 'the only difficulty with the words ... is the inconvenience they present' (cited in Sorrenson, 1989: 177). The Tribunal has promoted 'vigorous intellectual debate' that has fundamentally contested colonial assumptions and concomitantly exposed the 'historical amnesia that informs [Aotearoa] New Zealand's race relations myth' (Fleras and Spoonley, 1999: 21; see also Turner, 1999). Thus the work of the Tribunal illustrates the kind of agonistic pluralism Mouffe advocates, and demonstrates respect for the multiple criss-crossing and often contradictory trajectories that Massey argues coalesce in articulations of claims to identity and place.

Moreover, the Tribunal's work appears to be properly spatialised. Not only can it be seen as a 'moral authority in shaping the dynamics of [Aotearoa] New Zealand's social, cultural and political life' (Fleras and

Spoonley, 1999: 17), it creates the sense that te Tiriti is timeless and future oriented (Byrnes, 2006). The Tribunal has analysed 'the past' within contemporary conditions, providing for a timelessness, or 'out of time' character, to te Tiriti (ibid.). This sense is further reflected in the now accepted view of the Treaty as 'a living document' or a 'developing social contract', both by the Tribunal and across government departments, thus encompassing the unmeasured nature of Māori conceptions of time (Walker, 1982).

As a 'living document', it conjoins past and present, and can be interpreted according to contemporary conditions rather than just the conditions under which it was signed in 1840. The use of the phrase 'a living document' in government departments is further testament to the complexity and often contradictory nature of the place of the Treaty in Aotearoa. Te Tiriti, read through Massey's thinking, traverses multiple space-times and provides 'counterpoints in a state of continuous tension', opening up creative spaces of possibility (Fleras and Spoonley, 1999: 17). The Waitangi Tribunal, therefore, has been centrally important for Māori, not only because of the role it has played in facilitating the settlement of grievances, but because it has continued to challenge the linearity of time under colonial rule (Awatere, 1984: 61). As noted above, Māori conceptions of time are relational, inherently spatial and multiple – reflecting Massey's own re-theorisation of space-time. The Tribunal's work, then, as an agonistic space and a properly progressive politics, exposed and made room for Te Ao Māori in the public sphere. In Nancian terms, the Tribunal has provided a space of freedom, in which to expose (compear) iwi relationality (with past, present, future place and people) and their alterity. The act of remembering enabled through the Tribunal has affirmed a simultaneity of stories-so-far and alternative geographic possibilities to be imagined.

The closure of formal settlement

In contrast to the Tribunal's progressive agonistic space, the legal settlement process closes down a more progressive politics and potentially constrains future ways of thinking about being Pākehā and Māori. For example, the formal process privileges pan-Māori groups that are given a fixed identity as a particular 'tribe' or 'iwi' (Orange, 2004; Smith, 2007). The Crown will not settle with smaller, less formal groups even though, in the late twentieth century, many of these categories were contested and resignified, recognising the embedded dynamism of the terms for Māori ways of being (see Fleras and Spoonley, 1999; Smith, 2007). In addition, the dominant discourses associated with settlement have become focused on Māori development (often economic and reflecting a neoliberal corporate model), which serves as a distraction from the issue of Māori–Crown (and Pākehā) relations and redress for past wrongs (see Fleras and Spoonley, 1999: 133–139; Larner, 2002).

There are three points to note here. First, the settlement process has become based on restitution and claims making to explicitly achieve 'full and final' settlement in exchange for a package of compensation measures (Fleras and Spoonley, 1999). Consequently, it establishes opposing sides, in formalised conflict, scrambling for as much as they can. The claims-making process has encouraged a combative and closed politics of demand (Day, 2005), in which interest groups join together through an equivalential logic that subsumes smaller players by reducing the differences between them (Laclau and Mouffe, 2001 [1985]; Laclau, 2005).

Second, the claims-based model raises the stakes, and channels the focus to what can be gained/sacrificed in compensation. The settlement process creates 'an elite game of fiscal redistribution' (Fleras and Spoonley, 1999: 146) in which certain authorised Māori groups have rights to negotiate claims while others do not. In turn this creates a hierarchy amongst Māori, while concomitantly fixing identity so that that hierarchy becomes institutionalised. The corporate iwi model channels 'the Māori' problem, as it has been seen in assimilationist and integrationist practices in the past, into the hegemonic neoliberal system. Within the constraints established by the dominant partner, such a politics of demand can only ever provide opportunities for reform because it leaves no space to reconfigure or transform hegemonic relations (Day, 2005).

Third and relatedly, the constrained process is centred on full and final settlement by 2014. Not only does this mean that the Crown can potentially, and conveniently, remove the 'Māori problem' from their future agenda, but it means that there is no space in this linear trajectory for envisioning an alternative that may further the principle of partnership or meaningful engagement in Māori–Crown relationships (Fleras and Spoonley, 1999). Unintentionally, this raises the question of the embedded nature of colonial relations: if Māori were to exercise tino rangatiratanga (chieftainship and self-determination) as guaranteed in te Tiriti, it would 'fundamentally challenge the prevailing assumptions underlying white settler governance' (ibid.: 110).

Race relations in Aotearoa are therefore complicated and contradictory. Te Tiriti is reified and provides a means through which indigeneity and difference (to a degree) can be recognised. Yet this recognition is circumscribed by a model of social justice that is based on a notion that past wrongs can be ameliorated through compensation. The assumption within this model is that compensation will bring disadvantaged groups back onto a level playing field in which they can contribute to their own development as well as to the wider economy (Humpage and Fleras, 2001). Moreover, it allows the past to be 'put behind us', so that the country can move forward – a particular linear approach that again 'forgets' Māori conceptions of space-time or the 'simultaneity-of-stories-so-far' (Massey, 2005).

Such temporality annihilates the spatial, and is consistent with an 'imperial rationality' through which dominance is masked behind a shroud of benevolence and respect for difference (see Slater, this volume). It reflects the closure implicit in the post-political consensus in which the dominant Western linear trajectories become seemingly inevitable (Massey, 2005; Mouffe, 2005). It fails to fully acknowledge an ontology in which the past already lives in the present, or that multiplicity and diversity have been closed down by the assumptions inherent in the hegemonic history of (post) colonial rule, of which neoliberalism can be seen to be its latest manifestation (Bargh and Otter, 2009).

A Closing and an Opening: Towards a Progressive Politics of Place in Aotearoa New Zealand?

In this chapter, Massey's theorising about space and a progressive politics of place has helped us interrupt and rethink the myth that is known as Aotearoa's 'positive' race relations. By mobilising her understanding of space-time as the 'simultaneity of stories-so-far', we have been able to conceive of the Treaty and its associated institution the Waitangi Tribunal as potentially agonistic spaces, which can enable a progressive politics and, after Nancy, create appropriate conditions for the 'compearance' of iwi relationality and their alterity.

We have also been able to highlight how the potential and radical openness offered by the Tribunal has been closed down by the dominance of a claims-based, adversarial, neoliberalised, and still largely colonial process of formal settlement. The government's current requirement that all claims be settled by 2014 perpetuates a linear and closed temporality that annihilates the spatial, 'forgetting' the relational ontology at the heart of Te Ao Māori and in their struggles over place and practices *in* place.

Consequently, for us her work raises the question of how spaces for openness and a progressive politics of place can be created and maintained within the context of a post-settlement era Aotearoa New Zealand. Recent research in Aotearoa suggests one way (for example, Larner and Craig, 2005; Johnson, 2008; Bargh and Otter, 2009; Panelli and Larner, 2010). It shows how in certain spaces the past is understood as multiple and dynamic – it is always already in the present, rather than something to be compensated for and forgotten. Moreover, it occurs 'in place', and is therefore always already spatial. Inherent to the moments and relations that coalesce in these places, there is an understanding of the interwovenness of the genealogy of place through which identity is made and remade through social and physical histories (Bargh and Otter, 2009; also see Massey, 2004, 2005). Time, space and place are brought together, recognising the fluidity and processual nature of being, opening out to the very possibility of being differently.

Another way is to more closely consider our own positionings in relation to both openness and struggle. In a discussion forum with the University of Glasgow Human Geography Research Group in 2009, Massey discussed the role of academics and the shape of contemporary research practices and the push to make geography 'matter'. She suggested there may have been a shift since the 1970s when the motivation to do research 'came from real struggles that we were engaged in' (Massey et al., 2009: 407). She went on: 'one of the differences from the 1970s is that then we were the struggle. We didn't have to go and link up with the struggle. We *were* the struggle' (ibid.: 408, emphasis in original).

From our relatively privileged positions within academia in Aotearoa, we agree with Massey's sentiments. For us, as academics and as members of the dominant majority, we believe we have a responsibility to disrupt myths and to be political. We need to educate ourselves about alternative geographies and to embody struggle and a progressive politics within our places/spaces of academia. The colleagues mentioned above have done some inspiring work in this regard on Aotearoa New Zealand geographies, and we would like to see more. In particular, we would welcome work that focuses on an 'a-where-ness' (after Thrift and Massey, cited in Le Heron and Lewis 2007) of the specificity and simultaneity of Aotearoa New Zealand's stories-so-far, and the power-geometries that continue to render certain stories more visible than others, so that we might struggle to change them.

We also need to create spaces in which dominant myths such as neoliberalism, globalisation and colonialism can be interrupted and contested, and alternatives put forth. This, we suggest, requires recognition of the entanglement of myth (as those social constructions that shape spatial relations and collective identities) and identities that are simultaneously singular and relational (after Nancy, 1991); a radical politics of contestation (after Mouffe, 2005); and an understanding of space-time as the 'simultaneity of stories-so-far' that are articulated through, in and with a progressive politics of place (after Massey, 2005).

Finally, there is a need to take struggle back into the heart of academia to provoke debates that disrupt the linear trajectories and geometries of power that perpetuate (post)colonial power. To do so, we must take openness and struggle seriously. Massey's work is inspiring in this regard. True to the heart of radical geographical thinking, she links her anti-essentialist conceptualisations of space with wider political theories that provide much potential for imagining and realising alternative geographies.

Notes

1 In brief, key differences lie in what was ceded to the Crown and guaranteed to Māori in return. Article I of the Treaty in English cedes sovereignty to the Crown.

In the Māori version, Māori cede kāwanatanga (meaning governance or delegated authority). Article II in English provides Māori with guarantees of the protection and governance of their possessions and treasures. In Māori they were guaranteed tino rangatiratanga, more properly meaning chieftainship and self-determination (closer to the English term 'sovereignty') over their lands, estates, fisheries and taonga (treasures). Moreover, Māori as Tangata Whenua (people of the land) do not 'possess' land, but are connected to land through their creation myths and genealogy. Western notions of property and possession did not translate (see Kāwharu, 1989; Fleras and Spoonley, 1999; Orange, 2004).

2   For Māori, the past – mua – is also translated as 'in front', while the future – muri – is translated as 'behind' because it cannot be seen or known (Williams, 1971 cited in McKay and Walmsley, 2003: 90).

Chapter Fourteen
Doreen Massey: The Light Dances on the Water

Ash Amin and Nigel Thrift

Introduction

In this short contribution we want to reflect on the political stance that we believe Doreen Massey has stood for during her career, and consider what it might portend. Massey's pre-eminence in revolutionising the content and conduct of geographical thought since the late 1960s is well noted. She liberated regions from the strictures of regional science, she revealed the connections between spatial inequality/change and capitalist restructuring, she rearticulated place and space as relational entities, she placed class and gender at the heart of social and spatial concern, she made the case for a situated politics sensitive to the near and the far, she revealed the excitement of treating space as richly as time has been dealt with in the annals of Western thought, and she cleared the ground so as to enable cities to be grasped in all their complexity.

And all this with enviable clarity of expression, infectious enthusiasm and persuasive logic: how could anyone think otherwise after following her arguments? It is also well known that Doreen Massey does not theorise for the sake of it. She does so in order to cast new light on the familiar and, most importantly, to elucidate the politics of thinking the world in a certain way. She has played a prominent role in introducing critical thinking, ethical practice and political commitment into geography. She has challenged orthodoxies in the discipline, added new objects of concern, and brought in diverse

Spatial Politics: Essays for Doreen Massey, First Edition.
Edited by David Featherstone and Joe Painter.
© 2013 John Wiley & Sons, Ltd. Published 2013 by John Wiley & Sons, Ltd.

voices from the critical left, and has done so by adopting a 'disquisitional' rather than inquisitorial tone. She can be fierce in her scrutiny of an argument, her defence of a considered position and her condemnation of dogma, but it is a sturdiness born out of the commitment to open and spirited argument and to the innovations and understanding that come from it.

Doreen Massey's work has been political in another, more direct, sense. She is an active member of the democratic left and has taken up many causes. She has campaigned for women's rights, against the Iraq war, and for worker participation. She has signed up to many of the struggles and tactics of the World Social Forum. She is founding co-editor of the journal *Soundings*, which bridges academic left writing and political activism. She has been close to the Sandinistas in Nicaragua and to those pressing for local democracy in today's Venezuela. She has addressed countless political rallies around the world pressing for social and spatial justice, empower- ment and equality. She is a public intellectual who makes educational programmes, features on TV and the radio, and argues with politicians and opinion makers about the worth of the socialist and feminist cause. She was closely involved in the making of the Greater London Council under Ken Livingstone during the Thatcher years. She lives frugally and ethically. She cannot abide waste, opulence, injustice and oppression. A lot more can be said, but the point is clear. She is an engaged public intellectual.

Strikingly, Doreen Massey's political stance is principled but not orthodox, coherent but not dogmatic, critical but hopeful. It speaks to the dilemmas of left renewal in a troubled time of conservative ascendancy, left ossification, political closure and inventive churn in a world continually on the move (Judt, 2010). Her work has displayed a play between constancy and innovation, critique and proposition, and particularity and more general tracing of political contours that would serve the left well in reaffirming its future necessity and relevance. But this is not easy. We take up some of the dilemmas of cultivating a particular cultural habitat and ecology of practices in the rest of this contribution.

Habitat

In his magnificent book *Flora Britannica*, Richard Mabey (1996) argues that, though many of the traditions that linked plants, people and places have faded, still their spirit abides in the care and attention that so many people still give to their local patch. These vernacular relationships with the environment – names, ideas, practices – are our equivalent of what came to be known as folklore. They produce a kind of continuity in an environment which is always changing, a way of stabilising through talk and stance the changing lie of the land. They allow the story of chance and turbulence to be told as the art of cultivation and care, of the management of time.

But what is also striking about Mabey's book is that he well understands that the countryside is in constant flux. Entries are full of invaders that have become a staple part of the vernacular – migrants that have become locals, so to speak. The ecology, with all its patterns and continuities, is full of surprise arrivals and unpredictable developments, constantly passing under the radar of folklore, constantly altering the composition and dynamic of the landscape. There are echoes here of Massey's insistence on the city or region as not merely the juxtaposition of difference, but as a meeting place of many geographies and 'jostling, potentially conflicting trajectories' (Massey, 2007: 89), as a habitat of multiple presences, a site where the 'throwntogetherness' of things (Massey, 2005: 141–142) does the active work.

In Mabey, as in Massey, what we see above all is a recognition of just how much the landscape changes (Massey, 2005: 133). And this point can be generalised to many other domains of life. Very little stays as it is for long. This is an insight which finds its modern origin in the birth of evolutionary theory and the discovery of geology, but it now goes far beyond that into much shorter time spans. Just as evolution has been revealed as taking place both faster and more widely than had been thought, counted in thousands rather than millions of years (Cochran and Harpending, 2009), so even the cultures that we had thought of as slow-moving, even stable, are revealed as fragile things constantly having to reinvent themselves and often being extinguished when they are unable to do so.

Let us give just a few examples. Geologists now refer to the 'Anthropocene' to warn of how the Earth's natural cycle has been altered and shortened as a result of the millennial accumulations of human activity. A new geological record has opened up without precedent in the Earth's history (Chakrabarty, 2009). In turn, the differences of biodiversity have been flattened out and its volatilities intensified due to anthropomorphic intervention, the movement of species, germs and genes, and deliberate experimentation with the alchemy of life. Perhaps the most rapid and dramatic forms of change in our times are those linked to climate change, restlessly altering the fundamentals of life in all parts of the world as they are wrenched out of settled states of climatic pattern. All is up for grabs as the extremes of temperature, precipitation, air flow and water availability all kick in simultaneously. And then, finally, the volatilities of culture have long been noted, from accounts such as those of Marx, who traced the transitions from slavery or feudalism to capitalism or communism, to changes in the technical and social relations of production, to contemporary accounts of accelerating disruption related to capitalist surge, global instability, technological advancement and cultural confrontation.

In other words, the world is unruly and ultimately uncontrollable for any concerted length of time. Whatever we do, we will suffer upset – and for several reasons. First, new combinations of entities are always coming into existence. Second, there are many entities that we will never know about. Latour has called these entities 'dark matter' or 'plasma' (2005: 243–244). Third, human rationality is approximate. It only ever knows what it is doing in part. But, underlying so much social thought is the idea that everything can be brought into line some day. Socialism has always had this idea of order – and if order didn't work out naturally, it could always be made to through the agency of all manner of totalitarian machines – and so have the various varieties of back-to-the-future conservatism which insist on the power of tradition. In each case, there is a kind of aesthetic of definiteness, backed up by a deeply held belief that others really ought to think the same.

Ecology

So perhaps we need to think in a different way. Of a world in constant process. Of a world in which change is inevitable and in which the centre never holds for long, even when it looks as though it is (the USA or China). Of a world in which human projects can only ever unfold in uncertain and open-ended ways. Of a world, as a result, in which political projects must be able to constantly adapt and tack as new ideas go for a walk (Hallward, 2009). This requires knowing that we stand constantly on the cusp, in danger of dropping the ball, needing to ensure that even the most long-lasting insti-tutions have to be constantly reinvented through a habitus which depends on improvisation to produce stability. Finding any kind of origin in this constant state of provisionality will always be an activity fraught with difficulty.

That seems to us to square much more with the provisional experience of doing actual politics, anyway. Things rarely go exactly as planned: unexpected events crop up, no one does exactly what they are told, programmes rarely unfold as billed on the packet. If we were to think in this way, perhaps it would be liberating. We might expect both less – and more – from politics. We would value the work of politics more – the systems, the organisation, the mundane tasks. We would celebrate the foot soldiers equally since everyone is allowed the role of thinker, rather than just a few theorists or a usually self-chosen vanguard (Rancière). We would let all kinds of flowers bloom through a natural tendency to debate, rather than adopting narrow doctrinal positions from which thinking is elaborated but never truly challenged. Doreen Massey may want to quiz us on whether pluralism in its own right offers a way through right and wrong, but her own writing on agonistic politics (e.g., Massey, 1995a) lends weight to the claim that without a natural tendency to debate, the tests of what is right and wrong are conveniently avoided.

If all this is so, then we are starting to talk about a politics of propensity, a politics which is always to some extent a part of an experimental field. It will roll along constantly with more or less intensity. There will never be a world where things will stop, where everything will have been done and dusted and everyone will turn to painting and gardening and truffle hunting. The natural challenge to this kind of view is to argue that it is either a form of wishy-washy liberalism or the betrayal of sacred principles – or both at once. These dangers exist, of course. But to close down things like this is to strip out the nuts and bolts of the political process, the adjustments and compromises that need to be made in taking things forward, in ensuring that goals are met, in delivering the promise to the nearest approximation, in wanting a social structure to be so overbearing, invariant and determinate that only its utter collapse can alter it.

Programmatic writing on politics tends to ignore the agency of the process, vain enough to think that the programme can be carried along on the shoulders of its principles and supporters as an unchanged icon. But politics, at least democratic politics, is played out – in many cases formed – in a dialectical field. It is the clash of interests and desires, the tactics of position and coalition, the play of forces pouring into the field that does the vital political work. The gains and losses, the achievements and failures, and often the programmes themselves, are formed in this zone of engagement. The process blurs the distinction between means and ends, plan and pragmatic adjustment.

Cultivation

Mabey's *Flora Britannica* also tells the tale of maintenance – in this case through the powers of folklore. So too it is with politics, and so too must it be with a politics of continual adjustment if it is to sustain gains and keep momentum. Only on rare occasions does the political process break decisively with its legacy, shake off the chains that link the present with the past, jettison tradition. Change is slow to come and occurs in small steps (at least in mature democracies, times of peace and conditions of limited environmental turbulence). This is due to the weight of forces of preservation (locked into folklore, institutional practices, cultural habits, normative stances and dynamics of evolutionary inheritance) but also to the accommodations and compromises built into the political process. It is exactly such re-centrings that frustrate the ambitions of radical political projects, which struggle to find grip by fostering new beginnings.

One option is to wait for the 'right' moment, the moment of catastrophic failure and hegemonic collapse. This is the moment of storming the Bastille, deposing the Tsar, invading the enemy territory, carrying through the coup d'état, marching on Rome or Berlin, sustaining the armed struggle, bringing down the state and institutional apparatus when the opportunity for victory presents itself. The politics of surge in these instances has certainly been effective and few would deny that the results have managed to break radically with the past. And some of them have promised more than a glimmer of hope for the masses or the subaltern during or shortly after the revolutionary moment. But it can also be a macho distraction, as in some of the work of Žižek which not only elevates violence to the level of a revolutionary act, as though all violence can be counted as the sign of a revolution taking place, but also uses violence as a means of suspending the temporality of politics so that everything becomes more, well, exciting. But as Rothenberg (2010: 172) points out;

> Žižek's preferred example of an act is not [Rosa] Parks' nonviolent refusal to take a seat at the back of the bus, an act which was prepared for in advance by political organising and a discourse of legitimation, and which results in further politicisation and ultimately changes in laws and social institutions. Instead he nominates the heroic action of the 'famous Jewish ballerina' who shot the Nazi guards for whom she was forced to dance, effectively committing suicide in the process. The difference between the two incidents is striking: a nonviolent act that achieves politicisation in Parks' case versus a violent one that does not.

Yet, in the vacuum created by extreme instability, institutional collapse and insurgent hardening, far too many moments of revolutionary change have yielded a vice-like state politics of maintenance of order that has crushed hope, dissent, liberty and potentiality – above all, of the many who expected to be lifted out of misery, exploitation, fear and insecurity by the revolution. There are important differences in intent, aspiration and shade of outcome between these instances, and it would be careless to deny them – especially the differences between fascist, state-socialist and communalist regimes. However, the parallels between injustices across regimes of totalitarian maintenance are striking (Arendt, 1979; Judt, 2007).

How, then, might it be possible to commit to a radical politics of change and at the same time remain principled? How is it possible to stay close to the democratic impulse and to the surprises of multiplicity without becoming a slave to opportunity? How might we produce concerted agency without yielding to the intentionalist dream in which everything is clear to the end point? These are questions that underlie much of Massey's interest in a democratic politics – in London, Nicaragua or Venezuela, in the labour or women's movement – that is able to extend the rights of the oppressed and exploited without disabling voice and dissent.

Ironically, the 'pragmatic' turn in a so-called post-ideological age has begun to yield this kind of politics, characterised by parties and governments of different hue daily changing course in order to stay in power or adjust to the public pulse. This is a politics of promises altered, principles sacrificed, ambitions lowered, messages air-brushed, causes sentimentalised, enemies dreamt up, and ethics held in abeyance in the name of staying close to the democratic impulse. It is not this kind of politics we have in mind, for it is largely self-serving and light on principle, seeing democracy as a joust between already formed interests and actors. A politics of propensity – the word we have invoked above – requires creating space for inclination and tendency (to follow the English definition), but also for thought and imagination (as the Latin etymology suggests). It works through projection and the ecosystem's own energies.

A politics of propensity cannot avoid the obligation to trace the principles and sentiments of the future desired. The left sometimes seems to have lost this capacity altogether, disarmed by the victories of market society, the fine grain of numerous regulatory changes, the vilification of its intellectual legacy, the seductions of conservative nostalgia, and the return of religious society. It struggles to name what it stands for, or finds it hard to know how to burnish its image so that it moves hearts and minds. The left, true to its history of desire for a fairer and freer society, needs to become much bolder about matters of solidarity, equality, justice, security, free association and ethical responsibility compromised by our times, as Doreen Massey's concern with spatial (in)justice and 'geographies of responsibility' suggests (Massey, 2004). This means returning once again to principles of allocation, ownership, regulation, distribution, authority and responsibility for the just and equal society. It means being precise about the impediments and the adversaries, doing more than naming capitalism, globalisation, empire or neoliberalism. It means tracing these principles through new highways and byways of development and change, such as ecological risk, transnational dependence, urban concentration, human displacement, technological society, geopolitical conflict, hyper-consumerism, corporatism, income polarisation, culture wars.

The left – howsoever defined – is caught up in many campaigns and it organises around many keywords, but it lacks a binding imaginary. It has often sought unity by sweeping the bads under a single name (e.g., capitalism, neoliberalism) and the goods under a single utopian diagram, both of which have sparked many a backlash. But what about a unity of sentiments and dispositions? Are there certain structures of feeling that the left can claim for itself and for the good society? There is no compelling or clear left emotional compass anymore, which cannot be explained alone by the victories of market, media, religious or conservative society. The will and the means to make intimate publics around new desires seem to have evaporated from the left. The repertoire of songs, stories, pamphlets, meetings, books, icons, celebrations, broadcasts, protests, workplace caucuses, educational and recreational ventures that gave tangible, emotional, feel to the objects of socialist, feminist, anti-racist, post-colonial desire, has thinned. So have the organisational forms and tactical practices to sustain dreams and gains, to keep the opposition in check, to show that victory is possible.

We can agree that the political climate is not conducive. But, there may be lessons to be learnt from the Obama campaign, from the ways in which web technologies manage to gather and mobilise, from the mass demonstrations against the Iraq war, from the resonances of creative forms of political mockery and satire, from the decentralised tactics of the World Social Forum, from even the diplomacy of the EU in areas such as environmental protection. These mobilisations weave together cause, affect and organisation in new ways, powered by strong sentiments of injury, hope and solidarity, yet affect too often seems to play no role in contemporary thinking on left renewal and unity even though much of the engine of politics consists of the surge that comes from structures of feeling attached to objects of political anger or hope that demand change. This is precisely why a politics of propensity requires an affective compass that kicks in before thought to combat and rail against the injuries of the unequal and divided society and to step up to a fair and democratic society.

The left must sound the bell for critical curiosity, for fairness, solidarity, for ethical responsibility, and the sense of the commons, and many other structures of feeling that can displace the aversions that the right has so effectively learned to master in recent decades, which have too often become the standards of political conduct, designed to hinder truth speaking to power, stifle democracy, encourage opacity, and privilege entrenched interests. The democratic left can occupy an eyrie which no one else has an interest in: it can sulk, convinced of its rectitude; it can become Machiavellian and gradually sink into the background of politics-as-usual – or it can press to open up the political process to more actors, more issues and more decision makers. We believe that the latter course is the only viable one, for it alone creates space for latent matters of concern to make their way into the public arena, and for the left cause – however defined – to fight its corner, explain the shortcomings of its adversaries, stay close to the interests and concerns that need defending, and work to persuade and thereby constitute a public that can never be taken forgranted. The left has to stand for a different way of doing politics, one where the process itself is able to judge whether what it has to offer makes sense and is desirable, fighting for an active and plural public sphere that includes many parliaments and modes of delivery (Connolly, 2010).

Of course, this is a politics which is without guarantees – a politics of winning the argument and wrestling for power, though not for its own sake. It is a move towards reanimating the political as process rather than procedure. It still depends on plans – often minutely detailed plans – for sustenance but it realises that the plans are provisional and will never be able to be downloaded exactly as is. Political circumstances will intervene and force their own logic onto the situation, but instead of always seeing them as a problem to be overcome, they can be seen as an opportunity to be harvested, not least because these circumstances allow voices to speak and forces to intone that might have been overlooked or written off when they do, in fact, have something to say. Even old lines of power, seemingly permanently arranged to counter left predilections, can be reworked so that they can mean something quite different. All can be grist to the world-making mill.

Conclusion

Can utopia be reimagined as an affective compass, rather than an end point or a distant promise? That is the question. Societies are complex and that means they are perpetually emergent. They cannot be tied down. And we're in them. The idea of a future in which difference is dispelled disappears. A socialist future will consist of a shifting set of places and populations and cultural categories which continue to be in process: a patchwork. There will be no climax or epiphany after which the promised land of zero disagreement is founded. We will continue to have to imagine our place in the world.

None of this is to suggest that something called 'socialism' cannot be attained but rather that it will never be an ordered order and that, accordingly, it is unlikely to be as we imagine it. The best analogy may be with a map. There are those on the left who still believe that it is possible to establish an ambition/order like that portrayed on the sixteenth-century Waldseemüller map: a god-like vision of the earth suddenly accessible to all, something akin to the fulfilment of a prophecy. But there is another side to that same map: a backdrop for revealing something new; a signpost to another world that keeps on pointing the way. There will be no stopping, no time when it will all have been founded and found out, no time when it will not be fractious. But this is not the same as saying there cannot be a kind of unruly peace.

But to get to the new place – which in reality is the ecology we have in front of us and the tools of cultivation that we choose to work with – we need to change how we view the world. We need to reorient. Just as in former times, East, not North, represented the origin of things where time had begun (*oriens*, the Latin root of orient, means 'rising') and was at the top of the map (Lester, 2009), so we need to find a new point of rising – against the established order and equally against its unyielding shackles, a new beginning to time, a new North. We believe that it is this kind of reorientation that distinguishes Doreen Massey's work, a new umbilical that reconnects the world but in radically different ways. She has wanted to make a different map of the world, one which institutes something called socialism, but realises that it will have a geography which allows difference to continue to thrive. We might see this as a kind of animism in which bodies can become the same but natures continue to differ.

Chapter Fifteen
Place, Space and Solidarity in Global Justice Networks

Andrew Cumbers and Paul Routledge

Introduction

Reconciling the complexities of a politics of place with a commitment to a broader transnational solidarity has defied leftists from the time of Marx onwards. It has also been a theme with which Doreen Massey has valiantly engaged throughout her career, and, more than any other geographer, made critical and decisive contributions to. From her earliest academic concerns with uncovering the structural foundations of uneven development evident in regional crises of the 1970s (e.g., Massey, 1978, 1979), to her more recent work on the broader spatial relations of power within which her home city of London is embedded, a guiding thread of Massey's work has been to remind people both of the broader responsibilities of places and, in turn, how places relate and are bound to each other.

If critical thinking on place and space has been the leitmotif of Massey's immense contribution to critical human geography, thinking and acting politically about space has similarly marked her wider and impressive commitment to public policy engagement. Massey's career, unlike many academic leftists, has been characterised by a remarkable degree of political activism (see Massey, 2008; Massey *et al.*, 2009), within and beyond the academy, from her early work with the Centre for Environmental Studies, to her engagement with the Conference of Social Economists, her membership of the Greater London Enterprise Board during the heady

Spatial Politics: Essays for Doreen Massey, First Edition.
Edited by David Featherstone and Joe Painter.
© 2013 John Wiley & Sons, Ltd. Published 2013 by John Wiley & Sons, Ltd.

early days of Ken Livingstone's Greater London Council, to a number of more recent interactions with regimes as diverse as New Labour in the late 1990s and Hugo Chávez's Bolivarian revolution of 2001.

For us, it is the fusing of these two strands – critical thinking about space and the development of a relational politics of place – that has been inspirational to our work on global justice networks and issues of transnational solidarity (Routledge, Cumbers and Nativel, 2007; Cumbers, Routledge and Nativel, 2008; Routledge and Cumbers, 2009). In particular, our concern has been with how broader geographical and political consciousness is forged between territorially defined and embedded social movements. Massey's work on relational understandings of place and politics has been critical for theorising transnational social movement practices in relation to broader discourses on the ambivalent and contested concept 'global civil society' (e.g., Anheier, Glasius and Kaldor, 2001; Keane, 2003; Amoore and Langley, 2003; Anheier and Katz, 2005; Routledge and Cumbers, 2009). Importantly, and unlike many others who write about globalisation, her accounts of extended spatialities are nuanced by a continuing awareness of the territorial politics of place and how this in turn produces globalisation. In her recent book *World City*, for example, she writes:

> the global is locally produced; and global forces are just as material, and real, as is the local embeddedness. Some local places are the seat of global forces. And in such a situation it may be the local place itself – what is stands for, what its identity depends upon – that must be challenged. (Massey, 2007: 21)

Massey's continuing ability to mesh the complex politics of place and the wider sets of spatial relations within which they are embedded has been instrumental in our attempts to conceptualise spatially extensive processes of solidarity construction.

Global Justice Networks, Flat Ontologies and Geographies of Responsibility

The emergence, increased incidence and coordination of protests and campaigns worldwide to contest neoliberal globalisation have arguably been the most significant development in transnational social relations since the end of the Cold War. In the wake of the 'end of history' pronouncements (Fukuyama, 1992) celebrating the collapse of the Soviet Union and the perceived victory of liberal democracy and market capitalism, the upsurge in grassroots mobilisations around the world in response to the continuing realities of global uneven development served as a rude awakening to neoliberal elites (Tormey, 2004).

We have conceived of this overarching process of social movements working together across geographical space as being materialised through global justice networks (GJNs) – overlapping, interacting, competing – and differentially placed and resourced networks that articulate demands for social, economic and environmental justice (Cumbers, Routledge and Nativel, 2008; Routledge, 2009; Routledge and Cumbers, 2009). GJNs are marked by their concrete trans-local and transnational connections and practices and, given their geographical spread and the diversity of movements that comprise them, they are more significant in terms of international oppositionist movements than the Socialist Internationals of the nineteenth and twentieth centuries or the new social movements and anti-war movements of the 1960s. Genuine global networks of solidarity have been established that link hitherto disparate local struggles to broader campaigns and agendas.

Much of the discourse about these processes has been over-celebratory and has evoked notions of a 'global movement' (e.g., Kaldor, 2003; Hardt and Negri, 2004; Mertes, 2004; della Porta *et al.*, 2006), although more recent work (and indeed subsequent geopolitical realities) has punctured some of the over-inflated claims about the strength and democratic character of contemporary transnational solidarity (e.g., Routledge and Cumbers, 2009). Drawing on post-structural perspectives on power and space, the network has become a key metaphor for describing the operation of the diverse movements that compromise the alter-globalisation mobilisations (see Cumbers, Routledge and Nativel, 2008, for an extended discussion). Its use depicts what are seen as flatter, dynamic and more fluid forms of economic and social organisation that are emerging to reflect the 'stretching out' of social relations under globalisation (Giddens, 1990; Castells, 1996; Melucci, 1996; Urry, 2004).

In contrast to the flat ontologies evident in these accounts, Massey's recent work (e.g., 2004, 2005, 2007) displays far more sophistication in getting to grips with the complex spatialities of global power relations. She continues to problematise the politics of space and place in ways that acknowledge the territoriality of place amidst a broader relational understanding of the local and the global. This is most evident in her account of the continuing spatial inequalities and connections that frame London's relationship with both the rest of the UK and the world (Massey, 2007). In this sense, we think Massey has made a more thoughtful contribution than many, linking post-structural insights in relation to power, identity and relationality to a narrative that remains both embedded in a political economy of uneven development and politically committed to challenging the political and economic inequalities between places. In particular, in her attempts to recast the politics of place by considering how political intervention might develop against global neoliberalism at the local scale, Massey (2004) uses the phrase 'geographies of responsibility' to make the point that, because places are relational, and social relations flow through them, connecting us up increasingly to 'distant others' in complex ways, we should think more about how our own actions and interventions locally could be politically important.

Certainly, we would argue that the potential for GJNs to develop a sustainable politics of international solidarity involves not just understanding the way that the local is enmeshed in wider spatial relations, but also, and perhaps more critically, assessing how the 'global' is invoked in struggles that take place locally. Moreover, different places clearly have different capacities for resistance to neoliberalism, reflecting the uneven power-geometries that Massey argues are part of the broader processes of uneven development:

> a local politics of place that took seriously the relational construction of space and place ... would understand that relational construction as highly differentiated from place to place through the vastly unequal disposition of resources. This is particularly true of capitalist globalisation. The mobilisation of resources into power relations between places is also highly differentiated and a local politics of place must take account of that. (2004: 13)

For us, Massey's work has been useful for thinking through the different ways that particular social movement participants within GJNs operate. There is an in-built assumption in much of the literature that alter-globalisation networks by themselves are implicitly progressive, but at an empirical level it is important to understand how the territorially based movements that engage in GJNs negotiate the issue of geographical responsibility. Of particular importance are the motives of locally based movements for becoming involved in GJNs, and how their everyday practices attempt to take account of, or are reflexive about, responsibilities to distant others. For example, in discussing the problems of grassroots movements in South Asia being more actively involved in GJNs such as People's Global Action (Asia), an activist in a South Indian farmers' movement commented:

> Movements in South Asia have a limited resource capacity to fully engage in global solidarity, things like time, money, language skills and computer skills. Hence most Indian movements are not really ready to fully participate in a global movement, to commit to it full time, or to fully involve and engage the grassroots in it. Most movements in India are leader based and many of these leaders have neither computer skills nor English language skills and thus they profess to be uninterested in global organising since they so not possess the necessary skills for it. Most folk who do global organising primarily like to travel and enjoy the benefits of conference hotels – they aren't serious about global solidarity. The language of many movement leaders is influenced by NGO discourse and not by the language of the grassroots. We need to return to the grassroots since most global work is too much in the air. (Interview, Kathmandu, Nepal, 2006; quoted in Routledge and Cumbers, 2009: 210)

Clearly, tensions and contradictions always arise from attempts to develop an effective and sustainable politics of international solidarity whilst sustaining an effective place-based politics (see also Harcourt, Brooke Wilson, Escobar and Rocheleau, this volume). Much ultimately depends on the ways in which international consciousness is fostered among grassroots activists of movements that participate in GJNs to the extent that a wider spatial imaginary becomes embedded in everyday actions.

Furthermore, whilst GJNs need to be considered in relational terms, thinking through how the local and global are mutually constituted (Latour, 1993; Massey, 2005), it is important to remember that most of the actors and movements that constitute networks derive their principal strength from acting at the local and national scales rather than the transnational or global (Sklair, 1995). Hence, for many grassroots activists, whether it is in peasant or indigenous people's movements, trade unionists or even consumer activists, it is their own locality and sense of community that remain the most important source of collective and individual identities (Harvey, 1996; Castells, 1997) – or what Valins (1999) refers to as the 'stubborn chunks' of territorialised identity. Indeed, it is this local diversity and differentiation that is often under threat from capitalist modernisation. As Ceceña has put it with regard to the Zapatista uprising: 'Territory comprises ancestors, knowledge, the use of plants, their evolution, the perception of the cosmos, customs and community and living history' (2004: 361).

More prosaically, the realities of making a living, social reproduction and links to family and community structures continue to embed movements in particular places. Also, even the most mobile of social movement actors are never completely disembedded from these sorts of place-based social relations. They represent what Sidney Tarrow (2005) terms 'rooted cosmopolitans' – that is, transnational activists who move physically and cognitively outside their origins; draw on, and are constrained by, domestic and international resources networks, and opportunities of the societies/places in which they live; and advance claims on behalf of external actors, against external opponents, or in favour of goals they hold in common with transnational allies. Whilst there may be varying relations of connection to distant others, the continuing reality that everyday life is meaningfully territorialised is difficult to escape. However, crucially, immediate issues of survival and livelihood can act as motivations for people to participate in broader transnational networks (see Routledge, Cumbers and Nativel, 2006).

Spatiality, Responsibility and Solidarity

More broadly, Massey's relational understandings of the politics of space and place are critical in understanding the ability of social movements to transcend the local and the particular in the pursuit of more spatially extensive politics (e.g., the Zaptistas), that is, how extra-local responsibilities to others become embedded in the day-to-day activities of place-based or territorial struggles. In other words, in the context of a globalising economy, to what extent can places and movements foster a broader international solidarity and social justice agenda rather than pursuing a more reactionary and defensive politics? In this light, Massey (2004) was critical of the actions of the left-leaning Mayor of London, Ken Livingstone, during his period in office. Whilst some of his actions did deploy a more positive geography of responsibility – notably the introduction of a congestion charge to restrict traffic flow into the city, contributing to global CO_2 reduction – more generally, his economic development strategy did little to challenge the hegemony of financial interests and the discourse of neoliberalism. In this respect, he pursued a profoundly irresponsible local politics by failing to acknowledge London as a particularly powerful control centre within the geography of global capitalism, where more progressive interventions can have global effects.

> I am trying to argue something different again: that one implication of the very inequality inherent within capitalist globalisation is that the local relation to the global will vary, and in consequence so will the coordinates of any potential local politics of challenging globalisation. Moreover 'challenging globalisation' might precisely in consequence mean challenging rather than defending, certain local places. (2004: 13)

At root here is the need to develop a more self-reflexive local politics that recognises global responsibility with 'distant others'.

Where this takes us in thinking spatially about the relations between local agency and broader social networks within movements is to conceive of them in both vertical and horizontal terms, with the key concerns being how and why territorially based movements become involved in networks in the first place, and how the convergence of differently scaled and placed actors in horizontal spatially dispersed networks are played out in practice. Invariably, it is the recognition of deeper sets of power relations, underpinning local struggles, that produce empathy and solidarity with distant others. Conversely, particular place-based relations of power, such as those of gender, caste and class, may vitiate against the construction of solidarities with others. As Massey has noted, the diverse politics of place are themselves uneven and empower some groups at the expense of others (Massey, 1991a).

In the same vein, we have found that particular actors are often dominant within GJNs due to their control of key political, economic and technological resources (see also Dicken et al., 2001). Moreover, different groups and individuals are placed in distinct (more or less powerful) ways in relation to the flows and interconnections involved in the functioning of resistance networks. Thus, while the working of networks involves the intermingling of geographic scales, contradictions and tensions remain – either tied to the differential discursive and material powers enjoyed by particular movements and the social and spatial contexts from which they themselves emerge, or in the placing of specific actors within particular networks.

For example, one of us (Routledge) was involved in facilitating a network of Asian social movements, People's Global Action (Asia), between 2001 and 2007. An integral part of this work involved holding meetings with movement activists in their various places of struggle, and discussing ways that solidarities might be forged between them and other movements in the network. During a visit to Bangladesh in summer 2002, a series of meetings were held on islands (*chars*) located in the Ganges delta that had been occupied by landless peasants and defended by them against landlords and their private armies. On the island of Charhadi, a member of the landless women's association explained:

> Landless peasants, armed with brooms and chilli powder, occupied four uninhabited *chars*. On Charhadi, people have resisted the landlord and their armed *goondas* [thugs]. Through a series of signals the communities are warned of impending attacks. Despite successfully remaining on the island for ten years, people still have no education or health care, and no flood shelters for their cattle when the river floods during the Monsoon. Since the occupation nearly one hundred, mostly children, have died. (Quoted in Routledge, 2008: 208)

Clearly, the voice of this activist, the multiple voices of the villagers of Charhadi, the connections of their community to others, to nature, and to various social movements, and the connections of these movements to networks all form part of the 'flow' of relations that comprise PGA Asia. However, the everyday life of a peasant woman in a poor village on an occupied island in the Ganges delta is very different, for example, to that of a male movement leader living in the movement headquarters in Dhaka and connected to the Internet, which in turn is different to those of a Western activist-academic temporarily visiting the island. There are pronounced differences in physical mobility across space, access to resources such as money and technology and so on. Such differences in the distribution of power are more than just relational effects within the network. Rather, there are specific causes of such power relations, not least the complex web of political, economic and cultural determinations attendant to rural Bangladesh that include, for example, the oppressive economic conditions (e.g., loans) that delimit everyday possibilities of peasant life; the local political power of landlords in the local area which necessitates the defence of occupied islands by the landless; and the gender relations that circumscribe women's mobility beyond their villages (Routledge, 2008).

Moreover, particular places and movements become empowered whilst others remain marginal within the operations of GJNs. A range of place-specific conditions enable or constrain movements in their capacity to organise their struggles and participate within GJNs. Place-specific economic conditions – particularly the availability and deployment of financial, human, organisational, political, informational or cultural resources – are crucial in movement mobilisations. Transnational alliances are facilitated when movements possess significant mobilisation capacities already underway; when they have the capacity for regular communication with other movements; and when each organisation's members take some responsibility for brokering bonds of solidarity (Bandy and Smith, 2005). In addition, the ability of movements to participate in transnational alliances is also shaped by the actions, policies, limitations and challenges posed by the governments of the states in which they are located (Burawoy *et al.*, 2000; Glassman, 2001). In these ways, networks are both influenced by, and replicate, the existing 'power-geometries' (Massey, 1999c) that characterise connections between places under economic globalisation.

Place, Territory and History in the Politics of Social Movements

Unlike many contemporaries (e.g., Amin and Thrift, 2005), Massey's work remains robust in its recognition of the way deeply sedimented historical social relations shape the contemporary politics of place. The global city that is London in the 2000s cannot be understood without recourse to the broader geographical struggles within which it was entangled during the 1980s, as competing visions and spatial interests sought to remake both the landscape of London and the broader UK political economy (Allen *et al.*, 1998; Massey, 2007). Yet, that struggle was also shaped by the wider geo-economics and geopolitics of London itself and its changing role as imperial and financial hub within a global economy.

In a very different context, Arturo Escobar (2008) has shown how place-based ethnicities articulated by the Processo de Comunidades Negras (Process of Black Communities, PCN) in Colombia's Pacific rainforest region, have been confronted by, and resisted, the colonisation and domination associated with neoliberal globalisation. In so doing he shows how this region continues to be shaped by national and international processes of capitalist appropriation and extraction, as well by the actions of social movements such as the PCN in the rainforest region, in national legislatures, and in more international campaigns. However, PCN activists themselves argue that whatever the dynamics of 'global' organising, the costs of capitalist development continue to be borne by 'locals', and hence their organising proceeds from the 'local' to the 'global'.

Such insights are critical to our understanding of the potentials of transnational solidarities. For example, the spatially extensive networks of solidarity that were constructed in support of the Zapatista struggle, termed 'international Zapatismo' (Olesen, 2005), emerged out of specific territorial and historical context: the protection of historical rights of indigenous communities in Mexico to communal land, which subsequently became articulated within a broader critique of, and resistance to, NAFTA (North American Free Trade Agreement). Indeed the term 'Zapatista' refers to the revolutionary leader Emilio Zapata, whose original movement as part of the Mexican Revolution in the first two decades of the twentieth century was rooted in historical claims on land (Baschet, 2005).

In a broader sense, the construction and nurturance of mutual solidarities between workers, peasants, indigenous people etc. are likely to be predicated upon the common experiences of place-based alienation and exploitation through the workings of capital (Harvey, 2003) rather than with a more pluralistic post-structural left politics (Amin and Thrift, 2005). For example, an activist in People's Global Action (PGA) Asia commented on the potential for forging solidarities between different movements within the network:

> There are diverse languages and many local languages within countries, thus it is difficult to coordinate movements. But also there is a big possibility to spread the PGA process, since most Asian countries are agrarian-based and people are victims of globalisation and there are many movements because of this. People have different cultures and languages but common enemies, so there are differences but also potentials. (Quoted in Routledge and Cumbers, 2009: 116)

Looking at the other most celebrated examples of resistance to globalisation and neoliberalism in Latin America – Venezuela (where Massey's concept of 'power-geometry' has gained powerful political traction among the Chavistas themselves) and Bolivia – they are, and see themselves as part of, a wider spatial emancipatory project. Perhaps over-hyped in terms of its emancipatory radical politics, they do reflect a broader and more fundamental politics, aimed at rolling back neoliberalism in Latin America and in engaging in an alternative set of agendas around social justice. However, they are also embedded in deeply rooted territorial social relations that can only be understood in relation to specific geographies and histories of colonial repression and failed state modernisation projects. Of course, these are not in any sense territorially bounded, for they represent past political encounters between the local and the global, in the form of generations of colonial oppressors (Featherstone 2005). For the indigenous movements of Bolivia, the government of Evo Morales represents a break with centuries of domination from 'the outside', whether by Spanish Conquistadors, or by state modernisers located in La Paz. Indeed, drawing upon some of Massey's earlier work, it is possible to depict neoliberal globalisation as just the latest layer in historically sedimented spatial divisions of labour within which particular places are entangled (Massey, 1984).

Conclusion

While writing this chapter, one of us (Cumbers) took part in a focus group with young people in Motherwell, an old industrial town in the west of Scotland (July 10), for a research project exploring the effects of globalisation on communities (Mackinnon *et al.*, 2011). It is the kind of place that was laid low by the working through of Massey's spatial divisions of labour in the 1980s and early 1990s, and, with the collapse of its manufacturing base and the closure of its steel plant, subject to wider processes of uneven development, from which it has never properly recovered. The young people interviewed were looking for work in a depressed local labour market that had been badly affected by the recent global economic downturn. Not only were there no jobs, but they had lost the means of hearing about new jobs in their increasing detachment from the labour market. The common refrain was that jobs were going to foreign workers, prepared to work below the minimum wage, and there was a broad feeling that work should go to 'local' workers first. However, the youngsters also recognised the role of employers in infringing national employment law and the role of the state and training services in providing little sanction or opportunity for them. When asked about where they might look for jobs, the frame of reference was inherently localised; even the nearest big city, Glasgow, was viewed as logistically impossible, given the kinds of low-wage service job available and the costs and difficulties of getting there. None of the ten young men and women involved in the focus group had experience or knowledge of working outside the local area. Theirs were geographies of disconnection, yet through an economy linked to broader global political and economic processes, as Massey would be the first to point out.

The problems of a local disconnectedness, in the face of an increasingly complex and spatially entangled regime of global capitalism, raise some important concerns for those espousing global justice networks and broader solidarities as a call to political action. How does a progressive politics of place and a wider geography of responsibility emerge here? Where should the focus of our efforts on the left be? It seems to us that answering these questions and doing something about them – which is ultimately what organising through social movements means – requires a recognition of just how complex and entangled, and even multi-scalar (Mackinnon, 2011) geographical social relations have become through globalisation. These are relational issues that political activists have to confront, in the sense of the broader global processes and webs of relations through which the economy operates; meaning that asymmetrical power relations cannot be clearly located in one place, but are dispersed and diffuse. Paradoxically, though, these are at the same time territorial questions, in the sense that for most people, everyday experience remains rooted in local material realities, even though this does not necessarily reduce social and collective identities to a regressive localism. In this sense, we need strategies that deal with the local realities of the dispossessed, yet speak to more universal values. Forms of progressive localism (Mackinnon *et al.*, 2011) – such as 'living wage campaigns' – are relevant here in the way they prosecute agendas of broader economic, social and cultural rights as a way of addressing locally experienced injustices. In doing so, they create forms of solidarity across class, gender and ethnicity, which begins to unify subaltern groups. Global justice networks represent the potential (as opposed to the current reality) for the coming together – through convergence spaces – of these progressive localisms.

It is our conviction here that the reconciling of a deeply embedded and territorialised politics of place – the geographies of the everyday (for most people) – with spatially extensive social relations is the most urgent task for the next generation of critical human geographers. From her earlier concerns with uneven development and spatial divisions of labour, to the identification of power-geometries in the shaping of globalisation processes, and more recently in invoking a relational sense of place and geographies of responsibility, Doreen Massey has fashioned some important tools for achieving this objective.

Chapter Sixteen
The Socialist Transformation of Venezuela: The Geographical Dimension of Political Strategy

Ricardo Menéndez

*Every means of production requires the execution of an ad hoc spatial artic-
ulation to maintain itself.*

Joan-Eugeni Sánchez

Every Society has its Space.

Sanguin

Introduction

Societies transform themselves in a dialectical relationship with their
geographical time-space. Links with the textures and fabric of the past,
the various roles and structures where the concepts and essence of
relationships are reproduced, the intermittent forging of contradictions
and needs, the marks of space and time etched on the faces of the people
defining their culture – all these have been often used as examples cited in
the battleground of alienation and to justify the continued existence and
viability of the norms of a system. Transformative time-space sees the
breakdown of the underlying logic of a model's order and of the concepts
essential to its order and structure. It is the dialectical unity of form and
content, of theory and praxis, the essence of a new social model.

Spatial Politics: Essays for Doreen Massey, First Edition.
Edited by David Featherstone and Joe Painter.
© 2013 John Wiley & Sons, Ltd. Published 2013 by John Wiley & Sons, Ltd.

The time-space of capitalism has been characterised by deep social and economic spatial divisions (Massey, 1984). They are the asymmetries of the model, the ultimate consequences and geographical synthesis of the social and spatial concentration of capital, the framework of inequality, the exponential and inertial sentence of social and economic segregation within the space. Central to the Bolivarian revolution has been an attempt to move beyond the spatial inequalities produced by capitalist uneven development. Doreen Massey's account of geometries of power has been integral to this process (Massey, 1993a, 2005). This chapter explores the implications of geography for the shaping of political strategy in the context of the 'popular nation state'.

The Geographical Time-Space as a Social Dimension

> Space is not a scientific object removed from ideology or politics; it has always been political and strategic. If space has an air of neutrality and indifference with regard to its contents and thus seems to be 'purely' formal, the epitome of rational abstraction, it is precisely because it has already been occupied and used, and has already been the focus of past processes whose traces are not always evident in the landscape. Space has been shaped and molded from historical and natural elements; but this has been a political process. Space is both political and ideological. It is a product literally filled with ideologies. (Lefebvre, 1976: 31)

Space is not a term that can be strictly defined; it is not an absolute Newtonian or inexorable temporal component. It is a subject for analysis in itself and an authentic dimension of society, when viewed under the heading of geographical time-space. Every society has its space – an ideological and indoctrinating geographical time-space. It is neither a receptacle nor a simple reflection of social relationships. It conditions and is conditioned. It conceptually expresses a society's model and gives expression to its tensions. It promotes and affects the development of processes, whether within the cyclico-ideological logic of the system or within its structural or transformational components.

Space is where social, political, cultural and productive relationships interact, are promoted and reflected. The current time-space is the result of the framework of the previous model. Without a new productive model, a solid new social foundation (organisational, conceptual and ethical) and a new time-space, a change to the political model runs the risk of being restricted to a mere takeover of power, an attack on the capitalist system but with the bases left intact. The People's State involves overturning the liberal thesis of the Mediating State. The construction of a new geometry of power and a state with a communal and cross-sectoral base are the political expression and the praxis of the concepts which have revolutionised contemporary academic study.

Traditional Approaches to the Spatial Theme

Space and geographical time-space

Within 'the sciences', and particularly in the field of geography, there is a danger of using conceptual references as simple common ground, either nouns or adjectives that may have varied uses but without precise definition. *Space* is undoubtedly one of the concepts most associated with this problem, and *geographical space* perhaps even more so. Space is often used as an attribute which is mentioned indiscriminately and with different meanings.

This can cause two particular problems. The first of these is a conception of society where space is just one more variable and not an intrinsic dimension of the same. The second is a problem associated with interpretative approaches that fail to interconnect different modes of analysis. Once this definition has been critically revised in areas such as geography, it then becomes possible to obtain some general grouping parameters.

In this way the following five axioms can be identified:

1   A new type of space emerges out of the relationship between people and the physical environment.
2   Spaces are unique and related to the level of development of the society.
3   A relationship exists between people and environment which varies depending on the location (in accordance with the nature of the time-space relationship from which it originates).
4   The scale defines the level of detail (the relevance of relationships, significance of variables, etc.)
5   All subjects studied in the fields of natural and social sciences, as well as the arts, have a spatio-temporal component.

On the basis of these axioms, it now becomes possible to summarise a critical evaluation of the major trends of thought within geography and the other branches of social science related to the study of territoriality. For schematic reasons, it is worth first emphasising three fundamental viewpoints (Hiernaux and Lindon, 1993).

Firstly, there is a view of space as a container. This viewpoint assumes that space is no more than a location for objects where only unidirectional relationships are possible and in which space itself occupies a merely passive role. In this way only simplified relationships between the elements can be studied, such as distance and its potential translation into elements such as time and costs. Secondly, there is an account of reflective space. In this case, one starts from the assumption that space is a reflection, an element subject to the workings of the economy and society in general without a geographical base of any type. Finally, space can be seen as a dimension of the social totality. Ledrut suggests that geographical space is an alveolar structure which mediates between nature and society. It is a space 'created by Man [*sic*] to be occupied by him, and as such has a real and material dimension, but is also incorporated within social relationships' (Hiernaux and Lindon, 1993: 102).

Additional elements that stand out in the analysis of the concept of space include the dialectical unity between purpose, content and form (Menéndez, 1991). This is particularly significant in the case of art and architecture. Wallerstein offers three major time-space concepts (1997a). He positions three key moments as constitutive of different 'potentially subversive' articulations of time-space (Wallerstein, 1997b: 71).

These are, firstly, cyclico-ideological time-space, which deals with the evaluation of the structures which regulate the workings of system cycles. The aim is to understand the system's structure and anatomy as well as the dynamic within which self-regulatory tolerance and rescue mechanisms are activated. Secondly, structural time-space is a concept which attempts to understand both social continuity and social change, within the parameters of which both interaction and conflict occur. This notion of time-space stems from the idea that historical systems are organic – in the sense that they present a genesis, a historical life, and finally an end. This end may correspond to a collapse or a transformation. Finally, there is what Wallerstein terms 'transformational time-space'. This is the brief, rare moment of fundamental change; the time of transition between one historic system and another. It only occurs when re-establishing the equilibrium of the previous system has ultimately become impossible.

Using the above conceptual scheme one can finally obtain a defined focus for the evaluation of the terms specific to geography as a science of space, and its application in relation to the study of territoriality. As a result, we have now established a series of considerations which should be taken into account in order to provide an approximate definition of the term 'space' and its respective division into variables and areas of study.

A more detailed study of the subject rightly belongs outside the scope of this essay; however, it is still worth emphasising the dialectical unity that exists between space and time, form and content, structures and working. As I have argued elsewhere, 'we can understand geographical time-space as a dimension of the social totality, the product of the relationship between the dynamic blueprint of the historical system and the properties specific to the uneven texture of the same. Within this concept texture is understood as the weight of the previous social order.' In this regard geographical time-space is 'an indivisible system. It is made up of elements of the natural environment and socio-economic and cultural conditions which all combine to form a dialectical whole, contextualised in time and independent from their individual properties. Geographical time-space is the result of the use of the material in a state of flux under the combined impact of the

elements mentioned above. It possesses form, content and movement ...'
(Menéndez, 2007: 64).

Space does not exist outside time, just as human beings cannot exist
outside society, unaffected by the dimensions of the same. We could even
assert that every society has its own space, with which it maintains
causal correlations. In fact the structure of the land is both an expression of
social forces and at the same time influences and conditions the development
of the same. This is explained by the forms of land use, concentration of
population and resources exploited. In other words, to successfully achieve
a revolutionary transformation of society in Venezuela, both the theoretical
basis of the constitution and the means of production must be firmly rooted
in our time-space.

Direct Democracy: An Alternative Model for Addressing the Problem

The concept of direct democracy stems from a set of ideological values
radically different from those of liberal democracy. This is through its very
raison d'être, on the one hand, and on the other hand because of the range
and sphere of influences which define and are affected by it. In the first case,
the matter translates into the very definition of citizenship and the rationale
of power, whilst in the second, its implications reconstruct the desire to dissect reality in a self-interested fashion in order to reconstitute the dialectical
unity of a given historical model.

One of the defining elements of liberal democracy has actually been to
neutralise democracy itself as a form of social organisation or force for change.
In its place a political solution is proposed, with various forms of regulation
to control social tensions, but maintaining real power in the hands of capital,
as well as the resulting forms of cultural hegemony. In this way, one of its
structural foundations is dependent on the individual's conception of themselves. The ideological basis of direct democracy still recognises the role of the
individual, but further emphasises their position as a social subject. Meanwhile,
liberal concepts and declarations of the supremacy of the individual actually
represent in real terms a secure means of preserving the cyclico-ideological
time-space components of the historical model. In reality, not all individual
or individual interests are given recognition, as they are overshadowed by
the framework of relationships ensuring the supremacy of those who control
the social model and who are granted a level of citizenship in proportion
to the capital which is in their hands. In this way, the 'freedom of the individual' becomes a false premise used to articulate a defence, as a rationale of
the system, of the hegemony of a minority controlling the majority, in the
name of freedom and the expectation of a supposed 'social advancement'.

Marx and Engels have largely been given credit for identifying the class structure within society and its subsequent tensions as a central component of history. However, in Marx's own words, this discovery was not his own, but rather that of the bourgeois philosophers who preceded him; what was new to Marxism was the way to find expression and confront these tensions.

The central theme involves the conflict generated by class tensions and the forms of domination and distribution of power within the society, including the subject of gender as covered particularly by Engels. The premise entails the development of improved forms of organisation which would allow us to address the class conflict on the basis of the construction of a society of 'equals', resolving the processes of appropriation of the bulk of the workforce by the social partnerships controlling the organisation of the methods of production at a given moment in history. The same social partnerships which derive from the ownership of capital, technology and the means of production, as well as their spatial distribution, constitute the framework of inequalities which undermine the real basis for any model of full democracy. Society has been divided into two large groups represented by the owners of capital and the paid workers with the resultant forms of relationships and forces of production. That is to say, a system of social differentiation has resulted which correlates to the exercise of citizenship, in a way that is even more marked when the theoretical barrier of the political sphere is exceeded.

Hegemony and Power

Like the victor after a battle, a single model of democracy has come to dominate our codes of values and spread its deracinated vision of the world in the name of a supposed freedom. As the self-proclaimed authority on such terms, the liberal model has appropriated words such as 'democracy', 'freedom' and 'progress' through its culture and propaganda in the conflict of ideologies to justify the logic of capital in order to sustain inequality and exploitation.

The *raison d'être* of the state, its components and organisation, the forms of election, the function and conception of authority, the public and private spheres – all are prisoners to the hegemony of capital. In this way, the victory and dominance of the elites is ensured and representative democracy assumed to be the only model possible. The state is the regulator, made up of authorities distinct from the people, the great mass who may eventually be consulted to delegate but not to exercise power.

In the face of the crisis of capital and the neoliberal vision of the state, participation has been severely eroded. Genuine power for change has been taken from it and we are left with a residue of functions which the state is content to hand out when new forms of organisation are required, in the interests of capital, to represent the 'voice of the people'. Neoliberal globalisation has, as Doreen Massey argues, functioned as yet another 'hegemonic discourse' in 'a long line of attempts to tame the spatial' (Massey, 2005: 99).

The debate has had its compatriots. In the name of 'freedom and democracy', all options which did not spring from the dominant order have been removed. New forms of slavery have developed out of alienation. Direct democracy has developed in parallel to this doctrine over more than a century and a half. Looked at in this way, participatory and protagonic democracy has held a significant position within the political debate (see also Wainwright, this volume).

Protagonic democracy is defined in Article 62 of the Constitution of the Bolivarian Republic of Venezuela as 'democracy characterised by the free and active participation of citizens in the formulation, exercise and control of public administration'. This refigures democracy as a proactive, rather than passive, political relation and is designed to move beyond what Hugo Chávez has described as the failure of 'mere representative democracy' in Latin America (see Robinson, 2008: 339–340).

This participatory and protagonic democracy has the following elements. Firstly, it is conceived and executed by citizens themselves. This bears on the true democratisation of the term, without the restrictive parameters of the 'citizenship of capital' or concrete expressions of social exclusion. It is the basic Aristotelian vision of citizenship as the power to exercise and administer public authority.

Secondly, there is the exercise of political power. This speaks to the mechanisms of election and participation. If power cannot be delegated, then how can democracy be exercised if not in a direct form? For a long time, liberal democracy has propagated the idea of separate spheres of society, private and public, with the state as a mediator. Citizens are theoretically represented by the state, giving their approval according to their political interests, but not involved in either the theory or practice of the state or government. In participatory and protagonic democracy the vision of the nation state is a systematic totality.

Thirdly, there is the democratisation of the composition and functions of the state. This asserts the significance of democratising political, social, cultural, economic and spatial power which shapes the nature and existence of a society. It is also necessary to interrogate the value criteria, the goals represented in the media, as in a blueprint suggested by Albert Camus. It is not possible to construct new models, or even a transition, with practices disassociated from ideological ends.

This theme has become the central issue for change to the current historical model. In the first place, it involves breaking away from the cyclico-ideological and structural time-space of the capitalist model, and in particular from the forms identified as the precepts of liberal democracy. In turn, the challenge requires new paradigms in order to adopt a model which addresses a number of dilemmas. These include such questions as: who should make up the state, and what are its functions? Who makes up the citizenry and how do they exercise their authority?

Geography as a Component of Revolutionary Strategy

What will the new articulation of space involve?
Will it be the result of what we all do together, or what we stop doing?
Joan Eugeni Sánchez

The reconceptualisation of society and the practices necessary for its transformation represent a landmark in terms of both capital and the social sphere. That is not to say that these terms are exclusive. The issue is who will sustain the system's logic and for whom? Who will control the workings and efficiency of the system to enable the existence of true political, economic, social, cultural and spatial democracy? That is to say, what will be the relationships needed for the exercise of citizenship, the functions of the state, the articulation of the dimensions of society and the productive and historical dynamics?

Responsibility for the practical and theoretical development of these new concepts and paradigms may be defined as follows. There are tensions between organisations charged with political administration and those that are concerned with territorial integration. The problem is not simply one of semantics. The organisation of territory is connected to the political, economic and historical aims of society. A nation state requires a conceptual correlation with its organisational forms as well as with the actors in the public sphere. By adding the term 'popular', not only do the accents change, but also the architecture of the system. The driving forces and the workings of the configuration are different. As a result certain key elements for development will become essential.

Contrary to the theories of Fordism and post-Fordism, and even the exploitative model of commercialism, the logic of capital has led to a proliferation of models of spatial segregation and articulation involving other stakeholders. The break-up of land and the philosophy of competitive rather than comparative benefits have led to a geohistory of fragmentation. The model to put forward must now include a vision of the totality of space, a synthesis of functional integration within the blueprint of the system. In this way, material and energy will assume ever more complex organisational forms, where every unit makes up part of the organic fabric of the nation state. This must now be the concept to apply to the dynamics of integration within Latin America.

Regionalisation criteria are adopted in line with the functional, organic and strategic conception of the land. In other words, geographical applications should be made according to the essential principles of a socialist society based on equality and true freedom, with land distributed on the grounds of needs and capacities, whether in terms of regions, sub-regions, towns, communes and communities, with each having its own individual historical, physical and functional identity. This necessitates the specialisation of roles within the overall framework.

One of the key challenges here is to reconfigure the relations of space and power within the working concept of the state. There are important histories of thinking about the relations between power, democracy and spatial forms. *Toparchy*, for example, is a form of communal self-government advocated by the nineteenth-century Venezuelan philosopher Simón Rodríguez. This provided an embryonic model for the revolutionary state and a starting point for participatory and protagonic democracy. The principle of territoriality in state government and organisation is now crucial to its existence, life and development. In terms of taxonomy, various forms of land planning classification have been suggested through a correlation between areas of land and the social stakeholders; in other words, the land will shape the form of government of a space.

The 'popular nation state' is a complete vision of this concept on all levels, with a nucleus which stems from the community. Its existence does not rest on pure noise and commotion. Rather, it must be the spatial reflection of the community and not an arbitrary bureaucratic act which denaturalises its political and social expression. On this will depend the functionality and strength of the construction of the social fabric, the organic and geographical foundation of the state.

This popular nation state depends on the following key principles and directions. Firstly, there are new economic principles. These include the logic of economic efficiency; the principles of competition versus complementarity; the comparative advantages, opportunities and restrictions; the endogenous dynamism; accessibility; the levels and depths of functionality; the economy of scale of integration and synchronisation; the principle of solidarity and the logic of social capital.

Secondly, there are the spatial dynamics of urban segregation and regional planning and the concept of territorial functionality and democracy. The socialist principle of giving to everyone according to their needs and capabilities presents a real challenge: how to respect the individual in the face of diversity and manage the dynamics of land income and use; and how to determine rights of access in towns, as well as urban infrastructure and services and the very dynamic of the city. We need a new model of economies of scale which does not require the same spatial concentration and can maximise benefit through a connectivity and architecture of networks.

Finally, the construction of the social fabric of the revolution involves creating new geometries of power. The various levels of regional influence and the planning of different working scales and new development will require a fresh strategy for the construction and solidifying of a radical cultural dynamic of power, in addition to replacing the organisational structure. A popular power network involves full interaction between its organisation and cultural base to ensure the balanced success of the social, economic and spatial fabric of the revolution.

When we talk of recognising cultural diversity, for example, we are also referring to the fact that in the new geometry of power the existing inputs will also become catalysts for change to the organisational dynamics. This is reflected in the redefining of the social purpose and ownership of land by its occupants. In this way, one of the main principles for the democratisation of space lies in moving away from the role of passive tenancy to becoming the inhabitants and protagonists in the changing dialectic of the land. Obviously if ownership and capital are not democratised, this will remain impossible – in the same way as if the decisions of the community and their capacity for resolution continue to be merely empty declarations.

As we have seen, a new institutional structure must have two simultaneous objectives: firstly, efficiency and effectiveness through participation and direct democracy with the integration of varying levels of authority; secondly, a support structure and social fabric for the construction of a new society. This leads to the following key questions and challenges. What form will the geographical time-space of Venezuelan socialism now take? What will be the spatial architectures of popular participation? How will the integration of decisions actually work, bearing in mind the diversity of communities? How will we break away from the inertia of past history and the domination and exploitation of our territory? We need a new hegemony, based on humanity and inclusivity under the direction of a popular and communal state, where individualism involves integration with the totality and vice versa. The crucial moment must come when as individuals we feel part of the collective, with a new consciousness and paradigms that are far greater, fairer and more beautiful.

In thinking these challenges, Doreen Massey's insights in relation to the construction and negotiation of geometries of power have been crucial. Her work has helped to foreground key challenges and political questions around regional inequality and the importance of negotiating these for a contemporary socialist project. In particular, her stress on the dynamic, uneven and contested spatial processes through which power-geometries are made and remade has been instructive. This political application of her concept has also deepened her own understandings of spatial politics (Massey, 2011c). She contends that her 'appreciation of the forms and nuances of power was vastly enriched; the concept itself was dynamised – precisely because it was being used in a political *process*; I have been provoked to think yet more about the potential relations between places and politics' (Massey, this volume, emphasis in original).

Understanding time-space as transformative and uneven requires us to accept a range of fresh ideas and political and cultural practices, and to take on the challenge of a new conceptual and practical framework for the revolutionary process. There will be a new geometry – the geometry of popular power.

Chapter Seventeen
Place Beyond Place and the Politics of 'Empowerment'

Hilary Wainwright

Introduction

Several themes recur in discussions with Doreen, relaxed in her favourite Greek restaurant or snatched after a meeting, in a late-night station cafe.

The first theme concerns the relationship between the local and the global, the way they interrelate, both in terms of how the global order is shaped – in and through localities as well as in and through global institutions – and the manner in which it is contested. I share Doreen's insistence on transcending local/global dichotomies and instead exploring the interconnections. Indeed, I would argue that locating the contradictions and vulnerabilities produced by these relationships is a key to bringing agency into the picture and refusing the pervasive presentation of corporate globalisation as natural, invincible.

We are both born and bred in the north of England – Manchester and Leeds – and this northern background is a lasting and positive influence. We both enjoy London – its intensity and its ethnic mix, its surprises and contradictions. We both travel incessantly to explore, share and learn. But we feel at home, too, campaigning for a local hospital or canvassing for a local candidate of a radical socialist bent.

Through Doreen's work – for example, her analysis of London as a locality that shapes the global in a particular way – I've understood the importance of a political understanding of geography. For instance, in the

Spatial Politics: Essays for Doreen Massey, First Edition.
Edited by David Featherstone and Joe Painter.
© 2013 John Wiley & Sons, Ltd. Published 2013 by John Wiley & Sons, Ltd.

case of London this means recognising the significance of London as 'a world city' other than in terms of finance: for example by focusing on its multi-ethnic character, which is in fact threatened by the impact of its financial role on the affordability of living in London.[1] More generally, it means seeing geography as providing the intellectual tools for understanding the contested nature of dependencies, interconnections and effects between the local and the global. Through her work, I became curious to follow more systematically the intuition that leads activists today to work at so many different geographical levels in our efforts to change the world. I am thinking here of the assumption that the global is made and can be unmade at a local level, and at the same time the local is shaped and can be reshaped through global power and counter-power.

In building now on Doreen's work, I want to focus attention here on identifying the processes of *unmaking* and *reshaping*. Doreen writes about the constantly contested way in which London has helped to produce and reproduce neoliberalism. I want to highlight the nature of the flows and connections in the uncertain, experimental process of making of an alternative, including the role of a variety of cities, for example Porto Alegre in Brazil and Cochabamba in Bolivia, in stimulating and inspiring this process. I will draw on my own and others' experience of the World Social Forum and the counter-flows it produced and reinforced, connecting the local to the global and inspiring the confidence to resist.

I will then return to London and reconsider the importance of Ken Livingstone's Greater London Council (GLC), a moment when there was that rare political breed: politicians who refused to defer to the City and give it what it wanted but instead illustrated that the London economy, and even more the national economy, need not be so dependent on finance as a distinct 'sector'. Developing and implementing this commitment through the London Industrial Strategy[2] (LIS) was simultaneously an exciting and difficult experiment in which Doreen and I were close *compañeros*. Doreen has continued ever since to be engaged in politically influential debates about the future of London, consistently developing the case that the dominance of finance capital damaged London and the rest of the country, including their prospects of sustained and sustainable growth. Though wilfully and damagingly denigrated in the political memory, this example of 'modern socialism', in the words of Norman Tebbit as he vowed (precisely for this reason) to kill it, demonstrated that market-led politics was not the only way to overcome the limits of the post-war social democratic settlement.

In the last year of the GLC, Doreen and I were involved, as socialists and as feminists, in organising support for the mining communities, through the wide networks of Women Against Pit Closures – another experience of place beyond place. Remembering this in the context of the Thatcher governments' abolition of the government of London (in most countries, the destruction of a level of government, against the will of the majority of citizens, is associated with authoritarianism verging on dictatorship) reminds me of the importance of class. Here, I mean the exercise of class power in the making of neoliberalism (Harvey, 2006b) – and the way this was followed in the 1990s by the construction of a political project, New Labour, based on a determined, near obsessional bleaching of class, portrayed as 'old-fashioned', out of legitimate political debate. Against this I would highlight the remaking and radical widening of understandings of class – including the gendered nature of class – that was taking place through the GLC; in the support movement for the mining communities; through rural movements, especially in the global South around landlessness and urban movements around control over city budgets. Here too, in this insistence on both the importance of class and the importance of opening up our understanding of it, Doreen and I would concur and collaborate (Massey and Wainwright, 1985).

Place Beyond Place

Awareness of the flows and connections taking human relationships and consciousness beyond place, yet to place, from global to local and back in surprising ways, struck me vividly as I left the south Brazilian city of Porto Alegre in January 2001, on the return journey from the first World Social Forum (WSF).

At the time, I was researching one of the New Labour government's micro projects of 'community-led' regeneration in east Manchester, near where I lived. Every Wednesday night, I hung out with an improvised youth club created by parents in a local school to provide football, snooker and a friendly space for local youth who at the time were getting their kicks out of joy riding with stolen cars and other such dodgy activities. The council's youth service was one of the first services to have been emasculated under Margaret Thatcher.

East Manchester had been the powerhouse driving one of the UK's leading industrial cities. Its empty factories and boarded-up houses bore witness to the force of the deregulated finance unleashed by the Thatcher and Reagan governments. New Labour thought it could continue the neoliberal project but contain the growing social dislocation with micro interventions that put the problem-solving burden on community self-help, backed up by local authorities – which meanwhile were being deprived of any funding or powers for council housing.

Here, near Manchester's city centre, was also occurring the transformation of a place that had been abandoned in the 1970s and 1980s by the flows and circuits of capital accumulation as profits from engineering declined. It was hoped that it would re-enter those circuits by way of the property market, then in the heyday of financialisation, albeit with a very different 'population mix'.

Manchester City Council believed that its best option was to use its remaining leverage, through ownership of land and social management of the population, to play the market to maximise the benefit, as it believed, for the city. This included attracting businesses to east Manchester on the back of schemes such as hosting the Commonwealth Games and attracting inward investment from companies like Asda Walmart. The commercial nature of the strategy, along with the somewhat paternalist mentality of the council leadership, meant that there was very little scope for an active, self-determining role for local organisations – or at least they had to press very hard to have any genuine control. Indeed, the more independent-minded activists were seen by some council officials as more of a nuisance than as critical allies and a source of bargaining power in the council's dealings with developers.

As I left Porto Alegre on the airport bus I thought of the young people in the east Manchester youth club, and also of the trade unionists and residents of Newcastle then resisting the way the global market was reshaping their city, restricting the opportunity to live on the beautiful banks of the Tyne to those with the money to buy. And I vowed that if I came to the 2002 WSF, I would make sure it was with activists from these two cities.

Others returning from the WSF on the same bus talked in a similar way. I remember African participants discussing how arbitrary it was that it was they who were present at this first Forum, how they would report back and make sure that, next time, there were delegates from local movements. In other words, here were the first connections in a process of different, alternative and potentially counter, flows and relationships that might remake both the local and the global.

The next year, with the help of Unison and progressive trusts, two leading activists from Manchester and Newcastle, Sue Bowen and Kenny Bell, joined me at the WSF. Now, ten years on, they both say that their lives were changed by the experience of spending five days with 150,000 strangers who found shared goals and common purpose in a city openly proud of its radical participatory politics.

Flows and Counter-flows

The flows that connected Porto Alegre to the rest of the world through 150,000 activists such as these were not secret, exclusive flows of finance, trade deals, licences and patents, but open flows of consciousness, knowledge, political self-confidence, feasible models and strategies, and organisational bonds. They were flows that became sources of both global bargaining power – in resisting the WTO – and local alternatives, such as the spread of genuinely participatory budgeting.

Reflect for a moment on the significance of these global flows for the capacity of local initiatives to challenge and remake the global, or for the emergence of new global actors such as the Our World is Not for Sale campaign alliance.[3] Take, for example, the flow of political self-confidence inspired by the energy and seriousness of the WSF. This had – and still has – a special significance for activists in the UK, after 20 years of market-driven politics and the defeat of most attempts at resistance. This market-driven politics was accompanied by the marginalisation of those who insisted that there was an alternative, as 'old' was bypassed by a historical process which was said to have reached its apogee in the achievement of the global market-place and the redundancy of politics. The political self-confidence to pick up the pieces, reach out, make new alliances and develop positive alternatives was constantly squashed.

Such self-confidence matters hugely to our ability to gather together the human and other resources to make things happen. It matters as much as programme and policy. To comprehend this, one only has to look at the other side and take note of the importance of the public school system for the distinctive ability of the British establishment to reproduce itself. Its flexibility, its ability to make and dominate new alliances, comes from instilling in every generation of this ruling class an extraordinary sense of 'natural' leadership.

In this context, while anger and a sense of injustice had been growing in the last decades of the twentieth century – it was this, after all, that led to the defeat of the Tories in 1997 – the confidence that there could be an alternative, never mind the self-belief to create one, was, in England at any rate, at the time of the first WSF, at an all-time low.

On her return from the WSF, Sue Bowen led an initiative to create a peace garden carrying permanently, in stone, the Forum's central message that 'Another World Is Possible', alongside a mosaic of messages of international solidarity. It is in a prime position, next to the site of the Commonwealth Games, now the ground of Manchester City Football Club. Kenny Bell, inspired by the participatory institutions of Porto Alegre, led a successful struggle not only against the privatisation of Newcastle Council's IT and related 'back office' services but for a democracy-driven process of public service reform. It has become a model for resisting privatisation and demonstrating a feasible alternative.

Multiply these kinds of micro changes by 150,000 and you can see the significance of the WSF – not so much as a stand-alone global organisation but as a new kind of convergence of located actors whose activities and strategies are distributed and interconnected, coordinated to varying degrees but not through a single centre. In this sense, the much-debated notion of 'horizontality' is not so much a matter of anarchism versus Leninism (though insights can be drawn from classic debates); rather it is an improvised necessity for organising, gathering intelligence and developing strategy to act at all the levels necessary to reshape a multi-polar, multi-scalar world. The concept of 'horizontality' is probably too mechanical for the organic and messy products of social interaction across national boundaries. And it doesn't take account of the hybridity of the 'alter-globalisation' movement and the networks stimulated by the World Social Forum – trade unions alongside squatters, political parties in tense but also supportive relations with militant social movements. But some conceptual framework is needed to understand the emergence of forms of coordination and convergence that are not unified through a single centre. Radical geography will be important in this since these organisational innovations are in part a product of the necessity to organise beyond the national state, to follow but challenge the flows of global capital.

The result is not only the many initiatives of people like Kenny Bell and Sue Bowen. It is also sustained global networks, new kinds of alliances with some success in reversing processes of corporate globalisation (Pleyers, 2010). A good example here is the movement, local, national and global, that is both resisting concerted attempts to privatise water and demonstrating, through exemplary local experiences, new democratic principles for organising public services.

Recombining Local and Global: Demonstrating that There Is an Alternative

The story of the 'water wars' began in the mid-1990s with concerted attempts by multinational corporations to privatise water facilities across the world. A symbolic high point came two decades later when, in 2010, Bolivia won UN support for the notion of water as a human right. This does not in itself put an end to privatisation, but it gives important legitimacy to the idea of water as a common good rather than a source of corporate profit. The struggle continues but the story so far, with its relationships between local and global, resistance and alternatives, research and action, gives a revealing insight into the unmaking of corporate globalisation and the making of an alternative.

A turning point came in 2000 in the dramatic resistance of the people of Cochabamba. La Coordinadora de Defensa del Agua y de la Vida (Coalition in Defence of Water and Life) succeeded in sending away the Californian multinational company Bechtel. Here was a clear demonstration that the seemingly overwhelming pressures to privatise were human-created power relations that could be humanly resisted and remade.

The story moved on through the creation of a global network, Reclaiming Public Water, initiated by the international organisation of public service unions, Public Services International (PSI), supported by the Transnational Institute and involving both local campaigns and national alliances. It has been the story of the formation of a global counter-power whose strength lies in its local and national campaigns.

Reclaiming Public Water has developed a strong sense of itself as a protagonist through its encounters with the private water companies at the Global Water Forum. Witnessing their global coordination face to face has been an important stimulus to the cohesion of the network defending public water. Research by the Public Service International Research Unit (PSIRU), mapping out both the global nature of the threat and what this means locally and nationally, has provided a vital foundation of the network, while having the confidence that there is an alternative has given it momentum.

The Reclaiming Public Water network has built up a process of exchanging information about effective public management that is now having a cumulative impact. Books consolidating this information through case studies from across the world have been translated into at least 14 different languages (Brennan *et al.*, 2007). A similar process is described in many of the case studies: trade unions and citizens coming together to defend public water, making contact and getting support from the PSI and PSIRU, finding the public utility often in dire need of reform, investigating the reasons for the failings of the public company, and putting forward and campaigning for alternative plans for reform. Luis Issara from the Federation of Unions of Water Workers of Peru (FENTAP), described the result of such trade union and civic-driven processes of reform as 'modernisation without privatisation' (Wainwright, 2011).

The Struggle over Alternative Futures – a Local yet Global Experience

Luis Issara was occupying the official rhetoric and subverting it. Beyond polemical purposes the language of modernisation is problematic – implying one future, one direction, albeit reached by different routes. But his phrase does point to the way in which the war over the future of water has been part of a more general struggle with private capital over the direction of society during recent decades. The fight over the future of London through Ken Livingstone's GLC was an early battle in this continuing conflict.

Looking back, the abolition and attempt to defend the GLC was a conflict over what political and economic relationship should follow the social democratic settlement of the post-war years, which had already been eroded by stagflation and undermined by the conditions of an IMF loan.

As Doreen insists, Margaret Thatcher's victory was a class victory, although involving a reorganisation of class. But a class victory over what? Norman Tebbit was serious and accurate when he referred to their murder victim as 'modern socialism'. In other words, the government's determination to see the end of a powerful and challenging government of the capital was not just a scrap over individual politicians or institutions but over a possible alternative for London, and by implication an alternative influence on the whole of the UK, and maybe wider.

Ken Livingstone's leadership of the GLC inherited a political institution founded on principles of redistribution. As a London-wide body with a London-wide tax base, it made possible policies that redistributed resources from, for example, the wealthy inhabitants of Kensington, Chelsea and Hampstead to the poor of Walthamstow, Newham or Tower Hamlets. The socialism of Livingstone's GLC was not about ditching basic principles such as redistribution but about ways of expanding the needs on which public resources were spent and involving the public more directly in how their money was spent.

Ken Livingstone and the Labour group leadership that Doreen and I worked with – people such as Val Wise, Mike Ward and George Nicholson – were formed or significantly influenced by political experiences outside of party politics (like much of the innovation of this time, in economics as well as politics). Feminism was an important influence here; so too was the radical trade unionism organised around alternative industrial plans and the strategic and visionary community organising to promote the needs of Londoners against property speculators right up to and including the 1980s. These Labour Party activists who led the GLC in the early 1980s had a vision and understanding of their limits as politicians, rare among Labour councillors and MPs.

As a group with the project of winning control over the GLC, they had come together through the struggle to democratise the Labour Party, including to give political expression to the radical movements of the time. They had been part of the Bennite movement of the mid-1970s – an early phase in the fight over what was to follow the post-war order. They were determined to carry this struggle forward by winning control over the government of London. Their project was not to win control of the Labour Party in London as *the* instrument of social change in the capital city. On the contrary, learning from the limits of post-war Labour governments, their approach, often more implicit than stated, was to work 'in and against the state' (London Edinburgh Weekend Return, 1979), unlocking the resources of the state presently utilised in favour of private business; to be instead supportive of Londoners' own independent efforts at social change.

They started from a recognition of the limits of electoral power to achieve social justice without strong and autonomous movements in society, including in the economy. The strengthening of the power of these movements and initiatives to bring about change – a genuinely 'empowering state' – was an important part of GLC strategy and policy. In this sense, Livingstone's victory in 1981 was one of the first, and few, electoral victories for the new left.

I say 'new left' because this heterogeneous and at least initially united political grouping – and the people like Doreen and I whom they recruited in some role – did not come from nowhere. Now is not the moment for a lengthy discussion of the GLC. But to understand the significance of this moment in the struggle between different futures for London, the UK and, broadly speaking, progressive politics, it helps to locate the leadership of Livingstone's GLC in the context of a contested and unfinished transition that had its origins in the rebellions of the late 1960s and 1970s. The 1970s also saw the origins of the financialisation that Thatcher and Reagan championed and for which the abolition of the GLC became a necessity.

This historical context sets the scene for the particular struggle in London in the mid-1980s over the two alternatives: GLC versus Britain plc. The central dynamics of this struggle continue to this day. On a global level, we haven't been defeated. But we have been marginalised sufficiently for much of our language to be appropriated by the neoliberal right, hollowed out and the shells thrown back in our face, as they complete the marketisation of public services and the destruction of the means of redistribution begun by Mrs Thatcher.

We see the stolen language now in David Cameron's 'big society'. Cameron has invited us to 'join the government of Britain'. He pledges to put us in 'the driving seat, to take the decisions that affect the life of our families and our communities ... We'll give you the power, so you can take control.'

But control over what? His idea of the big society is pitched at minimising the power of the state, while doing nothing to give people the power to control the private, 'free' market and the inequalities it produces. For without economic democracy, decentralisation of political power will reinforce inequality, shifting power not 'from the state to working people', but to those who already have the money, social networks and time to make the system work for them.

In this way, Cameron, following Blair (with different linguistic twists, to echo an appropriate theme in each of their respective party traditions), has deployed the language of the new left. Thus he uses the language of empowerment and enabling to update and relegitimate the emasculation of what we would have seen as an emancipatory form of freedom[4] to a negative freedom for the capitalist class.

Both New Labour and Cameron's Conservative model for the future presume a dismantling of social democratic state institutions in favour of the dynamism of the capitalist market. In New Labour, the idea was that this would produce revenues that could be stealthily redistributed to maintain the old level of public services, albeit delivered increasingly by private businesses. In Cameron's case the logic is taken further, towards a fully commodified, US-style service, leaving a minimal social safety net and exhortations on civil society to rally round to meet the needs of the poor and needy.

Few on the left have any illusions that, in order to challenge this, the social democratic state of the 1950s and 1960s can, Humpty-Dumpty-like, be put together again. The critiques of the movements of the late 1960s and

1970s were prescient. Those state institutions of the post-war settlement, for all their achievements – especially in terms of redistribution and decommodification – were organised in a way that was too paternalistic, their mechanisms of representation too mediated and opaque to allow for genuine accountability. Their hierarchical and siloed forms of administration left these public institutions unable to adapt creatively to the growing expectations and capacities of the citizens who had benefited from their provisions. They lacked the political power to challenge the increasingly oligarchic nature of the capitalist market. Indeed, key actors in this economic oligarchy had become increasingly influential in the upper echelons of the state, resulting in the hollowing out of what political democracy existed. And the Labour Party was too monopolistic in its view of political power to open up to allies in the wider society to take on these growing powers of multinational capital.

The new political leadership of the GLC would broadly share this critique. Additionally, although the GLC had powers, they were strategic powers, not really powers for running anything, especially so far as the economy is concerned – which is my focus here. They included powers over land use, purchasing, certain strategic construction and some housing, plus a general power to act 'in the interests of some or all of the people of London'. This last power underpinned a proactive policy of grant giving to all kinds of independent campaigning, research and innovative civil society organisations, and it was the power under which the GLC created a public investment board, the Greater London Enterprise Board (GLEB),[5] to implement the GLC's democratically developed London Industrial Strategy.[6]

But there was no way that on their own that these powers could provide adequate instruments to achieve the objectives of decent jobs (including for women, ethnic minorities, young people and others treated as a marginal reserve labour force), to develop socially useful technology, introduce economic democracy, ensure the flourishing of inner-city communities and so on. The combination, therefore, of political vision and institutional necessity led the GLC's radical political leadership to use its powers always in collaboration with, and through strengthening, organisations in civil society supportive of the GLC's electoral mandate. In this way, they brought to the process of achieving it additional sources of power and knowledge.

Who Is Empowered?

This explicit kind of alliance with trade unions and social movements quite often came under fierce attack from the Tory press, as if the GLC had until that moment been somehow in a sphere of its own, not allied to any class or economic interest. I witnessed for myself the kind of economic alliances that we had to change to implement the Labour Party's electoral commitments. When I went to the 'showrooms' that were to be the offices of the Popular Planning Unit, the section of the industry and employment branch that I was to coordinate, I found boxes of leaflets advertising the benefits of London for foreign investment, including the fact that its work-force was one of the lowest paid in Europe. The showrooms had been the offices of a business support department of some kind under the previous Conservative regime at County Hall.

While the GLC under its previous Conservative political leadership took it for granted that its role was to condone and promote London as a low-wage economy – maximising profits was presumably the key dynamic in their implicit model of economic development – our conception of economic development entailed quite the opposite. Not only did we believe economic development must be underpinned by an equitable distribution of income and a dignified living wage, but we believed that overcoming the alienated nature of labour and realising instead the creativity of collaborative and skilled labour – whether paid and in the labour market, or unpaid in the community and home – was fundamental to a new economic model. This was the reason why the creation of alliances to empower the people of London, as workers, future workers and as citizens, service users and social consumers, was central to our economic strategy. This highly experimental process of economic empowerment had two main dimensions.

On the one hand, we used the powers of the GLC to *strengthen the transformative capacities of civil society*. As far as economic policy was concerned, this involved several strategies. Firstly, it involved strengthening the capacity of workplace and sector-wide trade unions to prepare alternative economic plans/bargaining strategies for their industries. This involved working with unions, researchers, institutions of trade union/workers' education, and paying for shop stewards to have time off to share and extend knowledge about the future possibilities for their industry. It involved making industrial democracy a condition of GLEB investments. It involved using purchasing power to insist on proper training, wages and organising rights. It also involved the creation of 'technology networks' of different kinds that put together academic scientists and technologists with individual innovators and designers and other technical staff working for industry as a resource to labour- or community-driven innovation.

Secondly this strengthening of civil society as part of an economic strategy also recognised the importance to the economy of work performed in the house, usually by women, unpaid and unrecognised. The GLC's London Industrial Strategy has a whole section entitled 'Domestic work and child-care', which ended with 14 detailed proposals for, amongst other actions, making more public resources available to share and collectivise the unpaid labour carried out in the home: childcare centres, launderettes, care for the elderly and more, all in collaboration with groups already taking initiatives to meet these needs.

Thirdly, it involved treating citizens as economic actors with a vision and knowledge about the future, including the economic future of their neighbourhoods, especially when they were threatened by economic forces over which they had no control. Thus the Popular Unit supported local groups in Docklands to develop an alternative plan for the Royal Docks as part of their and the GLC's resistance to what is now the City Airport.

At the same time, the GLC used its powers to *lift or overcome oppressive constraints* that prevented a transformative project or movement from flourishing. This, too, had both a labour and a neighbourhood dimension. The GLC's powers over land use meant that materially it could do more to remove or block constraints imposed on democratic initiatives concerning neighbourhoods and communities than major private-sector workplaces.

Classically, this involved prohibiting speculative or profit-led developments that would harm the interests of Londoners. A good example was its use of compulsory purchasing powers to block speculative office development in Waterloo, thereby enabling the plans of local community groups around Coin Street to flourish. There were many other examples of the same approach, under the GLC's 'community areas policy', based on a similar combination of disempowering speculators on inner-city land and empowering the democratic community alliances that had plans to develop the land in the interests of local people and usually Londoners more generally. As far as labour was concerned, the GLC had little power to directly remove or block the powers of the large corporate employers. But it could use its political legitimacy, its powers of research and investigation and its powers to support workers and community organisation to make public the social consequences of decisions to close or rationalise London factories, as it did over Ford when it tried to close a part of the Dagenham operation. At the same time, it could use its powers over land to limit the options by which Ford hoped to sell the land. As a result, it did not stop the closure but it strengthened the ability of the unions to win concessions in terms of numbers of jobs lost.

Since our focus here on the GLC is not on the detail (see Mackintosh and Wainwright, 1987) but on the way that the struggle over its future came to represent the clash between alternative directions out of the impasse of Keynesian economic policies, the important point in this context about the GLC's challenges to the strategies of multinationals in London is the principle. That is one of an elected government pursuing its electoral mandate in alliance with those who benefit from these goals, and in the process strengthening the capacities of these civil society actors to help bring about economic change. Essential here, for democracy, is the ability and willingness of the elected government in its contribution to such an alliance to use its powers, at whatever level, to challenge the unaccountable oligopolies of the capitalist market to block these possibilities.

A Different Kind of State

Here is a very different notion of the role of the state to that which guided much of the social democratic state. It is simultaneously more radical and more modest: more radical in that it is underpinned by a belief in the possibility of surpassing the capitalist market – including in favour of a democratic, socialised market (Elson, 2006); more modest in the recognition that no state institution could achieve this on their own as state bodies.

State powers are necessary but not sufficient for socialising and democratising economic decision making. Necessary too are 'inside actors' with the practical knowledge, the technical knowhow and the capacities for self-government – the workers, that is, at different levels, including hands-on levels of management. Necessary too, depending on the sphere of economic activity, are those who are involved in the users' or consumer side of the economy. This raises questions of how the organisation of companies and services has to change to involve users, consumers, local communities and other economic decision makers. The increased urgency (or awareness of the urgency) of environmental problems has reinforced the need for this wider understanding of economics and this closer integration of politics and economics.

Enabling the Market versus Strengthening Democratic Control

It is already clear how the experience of the GLC can help to sharpen an alternative understanding of empowerment in contrast to the upside down version that David Cameron uses to justify a thoroughgoing empowerment of capital. It also poses new challenges for us, if we are to fill in and expand on the outline of an alternative direction for government provided by the GLC.

First, the model held out by the experience of the GLC's economic strategies goes beyond a simplistic centralisation/decentralisation dichotomy and focuses on *who* is being empowered, in *what way* and *for what purpose*. The aim was to empower organised citizens to carry out plans that creatively fulfil goals for which the governing party was elected, thus using centralised powers to disempower property speculators pursuing interests that would shape inner-city London in ways that Londoners had firmly rejected. At the same time, the Livingstone GLC decentralised control of this land, within a negotiated framework, to democratic community trusts run by local citizens. In addition, the GLC provided resources of expertise and funds for the development of these plans.

Thus the GLC experience points to a model of state institutions that give up their monopoly on leading social change and share power both upwards and downwards to international and local state bodies, and outwards to actors in civil society on the basis of criteria based on the mandate of the elected government. (So, for example, the GLC's grants policies were very rigorously based on anti-racist, anti-sexist criteria.) This sharing of power has been based on new understandings of knowledge and of transformative power coming from both the practice of recent social movements, from 1968 onwards, to the alter-globalisation movements associated with the World Social Forum and from theoretical insights they have helped to inspire (see Massey, 2005; Wainwright, 1994).

The Challenges of the GLC's Unfinished Business

This unfinished experiment poses huge intellectual and practical challenges for developing an alternative model to market-led politics. There is space here only briefly to consider three.

First, the GLC's innovative, if ad hoc, combination of sharing power with organisations of civil society and at the same time strengthening state powers vis-à-vis oligarchic anti-democratic market forces, raises an interesting question of democracy. Clearly a hybrid form, combining participatory democracy and representative democracy, is necessary and is unevenly emerging across the world. But how do the different components articulate? What conditions are necessary for participatory democracy to involve a genuine sharing of power and, on the other hand, for stronger state powers to be accountable, including to the participatory processes?

A second challenge arises from the GLC's experiment in answering the claim of the neoliberal model that only the capitalist market can provide the best framework for innovation. The GLC tried several ways of supporting the creativity of labour, through the workshops with shop stewards in different industries having paid-for time off, the technology networks opening up academic institutions as resources for organisations working on alternatives, and the GLEB's insistence that a condition of its investment should be industrial democracy, with trade unions having power, including power to initiate plans for the enterprise. But we faced several problems.

First, there were the deep asymmetries in the division of labour and distribution of time that reproduced the treatment of labour as a hired hand and producer of abstract value, the source of profit, rather than knowing citizens, producing use value (Elson, 1979). It is important to think through the strategies for addressing this fundamental problem, building on the existing resistance to this alienation and diminishment of labour coming from employees, students, the unemployed and casualised workers (Wainwright, 2011).

The second problem is to explore what, given a rejection of the capitalist market as a false means of discipline and accountability, would be the alternative mechanisms by which production – including of services – can be disciplined democratically to provide what people want and need (bearing in mind, too, the needs of future generations). This raises huge issues, including, I would argue, the importance of distinguishing the capitalist market from other markets, and hence exploring the ways in which markets could be democratically framed.

This area of innovation has tended until recently to be ignored on the left, on the basis that in Marxist orthodoxy it is capital that develops the forces of production (socialism would take them over). It has been innovative trade unionists and technologists such as Mike Cooley (one of our colleagues at the GLC) who have insisted that there are choices in how the forces of production are developed and that the design of technology must be a terrain of struggle over alternatives, not a given (Wainwright and Elliott, 1982; Cooley, 1984).

Finally, what are the implications of the GLC/GLEB experiment for the organisations of civil society as agents of economic change? A distinctive feature of the GLC's economic strategy was that its assumptions about change, about the goals of economics and about productivity in relation to these goals, were grounded in recognising the role of organised labour, of community alliances, of women's organisations and many other social movement groups as economic actors with the capacity and interest to collaborate in reconstructing the London economy.

This was a systematic strategy involving, for example, two units of the GLC's Industry and Employment Branch (the Popular Planning Unit, PPU, and the Project Development Unit, PDU), which were responsible for supporting a whole infrastructure of research and organising centres working with community, trade union, women's and ethnic minority organisations on economic and employment issues. The PPU, for example, had budgets that paid for shop stewards to have time off and research backing to develop alternative strategies across industrial sectors and districts; the PDU was responsible for grant aiding independent research and campaigning centres to work with social movements on economic initiatives (e.g., the work on domestic work, referred to earlier). Moreover, the architectural and planning teams of the GLC made their skills available to community alliances to develop their alternative development plans.

In these ways, the GLC and GLEB, in theory even if it were not always possible in practice, took two steps beyond the limits of Keynesian economic orthodoxy. First, its London Industrial Strategy aimed to intervene on the supply side, in production – rather than just acting on levels of demand. Secondly, it worked to support and put to productive, public use the energies and capacities of working or would-be working and consuming Londoners; potentials that would otherwise lie undeveloped and unused.

Does this experience, occurring at the height of Mrs Thatcher's deindustrialisation of the UK, have any relevance in circumstances – the final outcome of the Thatcher period of deregulation – where the predominance of finance has reached its height and now its implosion, destroying the welfare state and public services as it crashes? Does the London Industrial Strategy, and especially the active economic role it gives to civil society, give us any clues as to how we counter the destructive role of financial markets?

In the past, finance capital – speculation and the making of money from money – has only been effectively constrained and its destructive logic suppressed in conditions of war and its immediate aftermath, or in conditions where there are powerful public banks, a strong public sector and extensive industrial democracy (for example, the Scandinavian countries, whose institutional arrangements were the product of a well-organised labour movement from the 1930s onwards). The decentralised system in Germany makes finance capital less rampant partly as a result of the regional democratic checks and balances constructed after the war over potential sources of overweening economic or political power.

In a timely and provocative manner, Maurice Glasman has opened up an important debate about how to constrain the power and in particular the mobility of capital, by which it moves from the real economy to speculation, in search of higher rates of return. He recommends a German-style social market economy as a means of achieving the necessary 'microdemocratic entanglement' through regional control of the banks, corporate governance based on partnership, a commitment to skilled labour and to mutual and cooperative forms of ownership. Now is not the moment for a full exploration of this model, but the question it raises for our discussion of the GLC and the problems posed by Doreen about the power of the City is this: how can 'democratic entanglement' or, more simply, democratic control be achieved in conditions where finance capital has become like a bloated god, feeding off decades of deference, deregulation and effective blackmail?

Glasman points to the potential power of citizens' organisation rooted in alliances of faith and community, with London Citizens as his model (Glasman, 2011). Though he does not refer to the GLC experience, a part of its strategy, as we've seen, was to work with and support similarly strategically minded community alliances. The power of alliances in Coin Street and in Docklands, for instance, in resisting property speculators – with some success – depended on their deep roots across these communities.

Glasman, however, is dismissive of the potential of trade unions in the UK to play a positive role in any strategy of democratic entanglement. He describes trade unions as exclusively 'oppositional'. After two decades in which political culture in the UK has shrivelled, in the shadow of News International, fed by a starvation diet of inaccurate information, he is reflecting a wider ignorance of traditions of trade union initiatives and bargaining strategies based on positive alternatives. As we have already mentioned, the GLC was generalising from exemplary campaigns by trade unionists mainly in the engineering industry, who responded to factory closures with a variety of alternative plans showing the uses to which their skills could be put.

This positive and strategic trade unionism is re-emerging after having had the wind taken out of it by the defeats of the 1980s. It is evident, for instance, in examples of resistance to outsourcing and privatisation, that is based on alternative, publicly driven strategies for public service reform, and reaches out to wider community/citizen alliances. The experience of the 'water wars' shows how the challenge of privatisation has stimulated unions across the world to act on the basis of the public nature of their work and reach out to fellow citizens to build campaigns to defend and improve a public good, rather than restricting their role to the defence of their own members' jobs and conditions (Wainwright, 2011). There have been similar experiences in the UK, especially in local government (Wainwright and Little, 2009). The significance of trade union objectives concerning the ethics and purpose of their members' work has also come to the fore in the aftermath of the exposure of the endemic corruption and degradation of news gathering in the 'news'papers of Rupert Murdoch. It is clear that News International's wiping out of the National Union of Journalists (NUJ) created the conditions for this degeneration to take place. The NUJ has generally made journalistic ethics central to its concerns. It is now reasserting itself in the current efforts to build a plural and ethical media against the pressures of finance and an oligopolistic capitalist market. Another interesting development is coming from UNITE, which is working with the Public and Civil Servants union on an alternative economic and industrial strategy.

What these and other developments around alternatives have in common with each other and with trade union involvement in the London Industrial Strategy is that they fuse politics and economics. They transcend the rigid division of labour between politics and economics that has often narrowed trade union bargaining strategies and industrial agendas, and weakened the political power of labour.

These developments are still amongst a minority. Their importance in any strategy of democratically embedding finance in the real economy is that they are a source of organised knowledge and power *within* production, driven by social need and use rather than the goal of accumulation. They are only one such civic actor. The growing social economy, for example, is another. And here too, the GLC illustrated the role that a state institution can play in supporting the infrastructure enabling such a part of the civic economy to flourish, embedding finance in its own way in the real economy.

In spite of their decline and defeats, the unions remain in many countries the most well-organised, well-resourced civil society organisations (other than organisations around religious faith). There are many signs that they are prepared to put these resources behind movements for economic change, in which they are one actor among many. This is part of a refining of strategies around class that recognise the nature of class power beyond the workplace, in the liberalisation of financial flows, in the commodification of common goods such as land and water, health, education and knowledge, and in the destruction of democratic institutions of government. It is involving a rethinking of the institutions of older movements like the unions and the cooperative movement, including their geography, as well as absorbing the insights of the movements emerging to address the challenges and opportunities of the new information and communication technologies (Berlinguer, 2010). Clearly, there is much urgent subject matter for more discussions in a favourite cafe with Doreen.

Notes

1 *World City*, review by William Podmore, 2008: www.amazon.co.uk/review/ R1ZJM1MQHOEOLA (last accessed 23 July 2012).

2 Published in 1985 by the Industry and Employment Branch of the Greater London Council, the LIS brought together the results of collaborative work between GLC researchers and trade union and community organisations in all the main economic sectors and strategic areas (e.g., Docklands) of London – including domestic work and childcare. Each chapter concludes with proposals for action by the GLC, always working with others.

3 Our World Is Not for Sale is a North–South alliance of a wide range of social and trade union movements, both urban and rural, developed through the WSF among other global processes and a key actor in blocking the WTO's efforts to spread deregulation and privatisation.

4 More than 'positive freedom'; negative freedom meaning 'freedom from': in Thatcher and Cameron's case, freedom from regulation.

5 On which Doreen was a very active board member.

6 The abolition of this power was Mrs Thatcher's first move towards abolition of the GLC. It was perhaps the most important enabling power used by the Livingstone GLC to carry out some of its most innovative and radical policies of empowerment.

Chapter Eighteen
'Stories So Far': A Conversation with Doreen Massey

Edited by David Featherstone, Sophie Bond and Joe Painter

What have been your key inspirations and influences?

Doreen Massey: Well, they aren't intellectual in the usual sense, perhaps. I have never quite worked like that. I have been asked this question before and I couldn't say that there have been particular people or authors that have been guiding lights in that sense. It has been more – and this may be in part a product of the particular generation I have been part of – that the stimulus, the reason for asking questions and the ways in which debates got framed, have come out of being part of political movements, whether that has been in the late '60s and the '70s with the emergence of Marxism, feminism, sexual liberation, being part of the GLC in the 1980s, or the kind of stuff that has happened more recently, and more generally an engagement with politics. So a lot of my key reference points have been urgent debates provoked by things like that. If there is one person that really influenced me early on, and this is a very strange person to cite, it is Louis Althusser.

In the late 1960s and early 1970s Marxism was very much on the agenda and I read a lot of Marx and Engels and took part in a lot of discussions, and yet found it very difficult to count myself as a Marxist. Even though at a gut level, in terms of wanting to think class politics, I knew I was 'on that side', reading Marx was not convincing me and I think a lot of that doubt was about the fact that we were, at that point, concentrating on early Marx and its intimations of a human nature, and as a feminist I couldn't buy it. So much of it was very essentialist about sexual divisions of labour and

Spatial Politics: Essays for Doreen Massey, First Edition.
Edited by David Featherstone and Joe Painter.
© 2013 John Wiley & Sons, Ltd. Published 2013 by John Wiley & Sons, Ltd.

'natural' divisions of labour. The heterosexual family was treated completely unproblematically. And so I found it difficult to buy into Marxism, even though I was strongly committed to issues of class.

And then along came Althusser. I actually learned Althusser in the strangest of places, which was the University of Pennsylvania in Philadelphia where I went to do an MA in mathematical economics and regional science. (I did that as a kind of 'you ought to know your enemy' kind of thing. And I think one ought to as well.) But you can do elective courses in US universities, so I did this course in French philosophy which was mainly on Althusser. There were two things about Althusser which utterly changed my view of life and of Marxism. The first was that – you know, before all your Derrida, before all of these arguments that we all know now so well – Althusser insisted that everything was always a product of what came before it. And if that is true then it seemed that nothing is given. So anything can be changed: his famous line, which I have repeated in a number of my works, is 'There is no point of departure'. And that absolutely – as a young woman who was trying to escape the norms, who felt she didn't conform to any of the given descriptions of 'woman', and who wanted a way of challenging them – that first entry into anti-essentialism, although I didn't know that term, none of us knew that term at that point, was utterly important.

The second thing: Althusser was in the French Communist Party (PCF) and was wrestling with the standard critique of Stalinism at that time, which put it all at the door of a person – Stalin – and Althusser was trying to say 'No, it is bigger than that, it is more structural than that, the critique has to go deeper'. And part of the way in which he constructed a deeper critique of the PCF's long-term political history, and therefore of the Moscow Communist Party, was to engage very critically with economic determinism. He began to put onto the agenda questions of ideology, questions of structures of society and ideological state apparatuses, which immediately appealed to feminists. And in the women's movement in the early 1970s I was in two reading groups, reading Marx on one set of nights, and Althusser and Balibar and people like that on other sets of nights – in each of them doing a critique of the other. And of course Althusser brought onto the agenda the later Marx. So I 'became a Marxist' at that point through that rather odd trajectory. So, that is how I think about influences.

What does Althusser's argument that 'there is no point of departure' open up for thinking about place? How does this relate to Deleuzean accounts which stress place as temporary stabilisation of flows?
Althusser was my awakening to anti-foundationalism, that's the point, and therefore it links directly into an anti-rootedness of place. How does this relate to Deleuzean accounts which stress place as a temporary stabilisation of flows? Well, first of all, Deleuze is much later, so Althusser happened to be the moment when I woke up to it. But of course it does resonate very

well with a Deleuzean approach. In particular the relational view of place is very much in accord with some notions of assemblage. And also with the idea of a positive construction built out of heterogeneous practices and relations, rather than analysing such things in terms of incompleteness or lack. I wasn't really aware of the Lacanian roots of the Althusserian period when I first came across it, and find myself now very much aligned with Deleuze and Guattari on their stance against negativity (I write about this in *For Space*). Their open approach also makes it easier to bring in the non-human.

But I'm now perhaps more wary of the 'all is process' view of the world, in so far as it's been translated from a reconceptualisation to almost a denial of the existence of 'things'. And in the meantime you have to deal with the entity. The body is all process, certainly, but we still have bodies.

When I first read John Holloway's book [*Change the World Without Taking Power*],[1] with its emphasis on fleeting relations, it took me back to a demo that I happened to be present at. I had been in Nicaragua working (in the first Sandinista period) and you had to come out of Nicaragua every 60 days to renew your visa. So I was in Guatemala City and there was a demonstration, and there was a banner which said, 'The Roman Empire fell so the American Empire will fall'. That is true. Everything is indeed in process and all things will pass, but it may take ages. And in the meantime you have to deal with the thing.

I do think there's been a move to a point where we don't think about the constellations of those things themselves, we go on so much about flow. I want to insist that the process argument is about reconceptualisation, not a refusal of recognition. So (back to place), on the left we don't have to abandon place as an arena for the construction of politics. What we have to do (and I have tried to do) is reconceptualise it processually.

Picking up on this, is there a danger of relational theories of space being seen too starkly against the traditional use of space and notions of space as fixed? Is there possibly a middle ground between relational and more traditional fixed views of space?

The article called 'A global sense of place'[2] was all about that relational view – places being open, as unbounded meeting places. But I didn't mean that place does not exist, I meant that it has to be rethought and reconceptualised. As I said, I find myself in disagreement with a certain strand of geography which argues that 'everything is now flow'. I don't think that is the case. This is also an argument with certain strands of political theory, particularly John Holloway and Hardt and Negri.[3] Hardt and Negri say that you can't have a concept of place which doesn't have boundaries. Therefore it is one of those old, modernist containments that we must abandon, because it divides global humanity, it divides the global multitude, and it preconstitutes identity. I think that is a classic counter-position, like striated space and smooth space, about which I am wary. There never was place that was a bounded container, nor indeed an identity of place that was not contested.

On the other hand, neither is there a non-striated smooth space. What we have to do is take responsibility for the striations that are inevitably made; take responsibility for the inevitable boundaries, the definitions, the categorisations, in the sense of politicising the terms of their construction. So yes, I would want to hang on very much to place, but to an utterly reconceptualised notion of place.

One of the contexts for 'global sense of place' was the dismissive attitude of some on the left to the local. What is the relevance of your account of place in relation to the current attempts to claim 'localism' for the right?

The left has an ambiguous attitude towards the local. There are some who do tend to be dismissive – to talk of things being 'only' local struggles and so forth. But there is also on the left a veneration of 'local people' (though why people should be restricted in this way to being 'local' has always puzzled me – it seems so condescending).

Anyway, as you say, the right is now reclaiming localism in the UK. And I think this did for a moment wrong-foot a strand of the left which had worked with an imagination of 'local people' as the bearers of authenticity and of independent engagement.

The main argument, of course – and one that we should be making very strongly – is that 'the local' and 'localism' in themselves have no political content, either progressive or reactionary. To believe they did would be to indulge in a form of spatial fetishism. The local, just like the global, is a construction through power-filled social relations, and its political content will depend on the nature of that construction. Then we must argue about the *form* of any particular localism.

Seamus Milne in the *Guardian*, just after the announcement of the Big Society,[4] joked that David Cameron must have been learning from Hugo Chávez, because Chávez too wants to strengthen the local level. But of course the two 'localisms' are completely different. In Venezuela it's about the building of grassroots popular power. It requires the investment of state resources (because without that all decentralisations can lead to further entrenchment of existing inequalities). In Venezuela the strategy is leading, especially, to the empowerment of the previously disenfranchised, especially women. In the UK the Big Society is about the opposite of all of those things. The political content of these two 'localisms' is entirely different.

Your early work was motivated by an interrogation of geographical and regional inequality?

Yes, it began with industrial location, actually, and the reason I went to Philadelphia to do the MA grew out of this. I was uneasy with the then dominant industrial location theory, much of which was based on neoclassical economics. I wanted to get to grips with it so I had to learn neoclassical economics. The paper 'Towards a critique of industrial location theory' came out of that, in a very early issue of *Antipode*.[5] *Antipode* was still a hopeful experiment then, an early voice of radical geography, alongside other voices such as the Conference of Socialist Economists, in which I was active, and *Capital and Class*, where I was on the early editorial board. Anyway, this investigation into geography and industrial structure led me towards questions of regional inequality.

Spatial Divisions of Labour is an expression of one of the things I have most taken from Marxism over the years, which is thinking relationally. One of the jobs that book was trying to do was to spatialise the relations of capitalist production. That is what it was about. And the other thing it was about, and maybe this is another influence, was being a northerner. In English terms, I'm from the North-West and have lived with, through and kind of in combat with regional inequality since my childhood, and so a lot of what I wanted to address was the terms in which regional inequality, 'uneven development' as we called it then, was thought about. Looking back now, it seems so obvious, one has to put oneself right back 30 or 40 years to note that it was quite a new thought at the time, some of the things we were saying. But to argue that you shouldn't just think about regions in terms of them being successes and failures through their own behaviour, so that they are to be blamed or congratulated on their economic failure or success, was then quite new. And to argue that regions were locked into systems of the spatial organisation of capitalist relations of production, and that it was through that that uneven development got produced, was at the time quite radical. And it is still something I go on about all the time. The *World City* book,[6] which is provoked by London, is precisely an argument about relationality and geographical inequality. Anyway, it was from *Spatial Divisions of Labour* that developed all the arguments about the crucial relation between space and power.

Looking back, there are perhaps some interesting shifts to reflect upon. *Spatial Divisions* was about trying to conceptualise the relational geographies of power within the capitalist structures of class. Its perspective was that of class. By the time of 'A global sense of place', the geography of social relations was being viewed more – or at least additionally – in terms of other forms of 'difference', most especially ethnicity. (In the work I've done on globalisation and on London both are present.) It's an interesting shift and reflects a more general move within geography and the social sciences away from class and towards, especially, hybridity, multiculturalism and so forth. Personally, I think it is time for that balance to be redressed.

What's more, I think that shift within the academy is a reflection of – maybe it even contributed to – one aspect of the shift between social settlements that took place in the '70s and '80s of the twentieth century: that is, the shift (in the UK and other parts of the 'Western' world) from a social democratic Keynesian hegemony to a neoliberal one. We know well the general lineaments of that shift, but one aspect that is not usually noted is the change in the nature of the articulation between equality and liberty (or democracy – which entails equality – and liberalism). Comparing the Keynesian days to today's settlement, equality has lost out to (an eviscerated form of) liberty. I've just written a piece that addresses this – for *Soundings* number 48[7] – and, again, I think it's a balance (or more accurately an articulation) that needs redressing. Equality needs to be put back more firmly on the political agenda.

What has been the significance of the concepts of 'hegemony' and of 'social settlements' to your work?
Yes ... I guess another set of influences has come from Gramsci, or from a Gramscian school of thought, especially around Stuart Hall, Chantal Mouffe, Ernesto Laclau and others. Decades ago (I think it was in the early '80s) we were all members of a group called 'The Hegemony Group' – another very challenging discussion forum. It was at that moment when concepts of modes of production and social formation were really being contested. It was related to that wider move to take culture and power more seriously – the whole notion of the *construction* of a society and of its common sense; the way different instances both had a degree of autonomy and intersected; and of course the possibility of those moments of conjunctural rupture when the balance of social forces may be put in question and changed. This was to be a really important approach to understanding Thatcherism, especially in the work of Stuart Hall, and at *Soundings* right now we are trying to engage in that kind of analysis again, about the current moment. Is this point of economic crisis a potential moment of real dislocation of the existing hegemony (see especially issues 44, 45 and 48)? And through all of this I have been trying to weave a thread about the relation between space and power, about the nature of space and the nature of place. The notion of hegemony, for instance, implies both place and a particular – contested – notion of place.

Your essay 'Flexible sexism'[8] argues that the 'potential contributions of feminism' to politics and academe 'have routinely been ignored'. As a feminist geographer, how do you see the gender politics of academe evolving?
The reason I wrote 'Flexible sexism' was partly just pure, visceral anger. There are different ways of doing feminist geography and my preferred tack has been not to study gender as such. Feminism has been about more than gender. Feminism in the 1970s was about a new society, a new way of being, a new way of organising things; it is a bigger political struggle. I want to have feminists everywhere, in nuclear physics, in geomorphology, in human geography ... studying everything *as a feminist*, not just studying women or

gender. So I was quite resistant to doing gender. That doesn't mean other people shouldn't. One of the things that reflects one of the points you were making is the number of times I get invited to places and they've heard of you, and they invite you, but they don't really know what you do. But because you are of the female persuasion they assume you do gender. I accepted a major prize in a major European country, and they said, 'We always ask our recipients to do a seminar and we thought, Professor Massey, you could do one on gender.' And I said 'Why?' And it was like, 'Well, you are a woman'. Extraordinary! And I said, 'Well, I will do it if, next time you award this prize to a man, you ask him to do a seminar on gender.' And so there was a complete backing off; I did it on something else. So one thing is that I think feminists ought to be everywhere, and doing feminist geography is about a lot more than studying gender specifically. It is an outlook on life and a way of doing things which is a lot more subversive than that. And the other reason for 'Flexible sexism' was that I think we ought to attack the citadels. These were the two big books of the day [*Postmodern Geographies* and *The Condition of Postmodernity*]; everybody thought they were wonderful, and in many ways they both were.[9] But they also seemed to me to be utterly, unconsciously, deeply, sexist. And I just wanted to attack where the sources of power were in terms of the voices within geography; so I did.[10]

As to what we address now, yes, I agree with what you say. I have been lucky in that the Open University is pretty progressive. But in universities and academic institutions more generally there remains a pervasive oblivion to issues of sexism. The discounting of women, simply not seeing or hearing us, still goes on. But there is also, more fundamentally, the whole structuring of things. So, for instance, when you do join in on 'important' committees you still (after all this time and in order to be 'taken seriously') so often have to act in a way (pompous, overly serious, self-congratulatory, competitive – though in a very 'civilised' way!) that is counter to everything feminism should stand for.

There is also one issue that has been troubling me just recently. And that is how we as – and this would include men as well – anti-sexist, feminist geographers, academics, should be addressing issues outside of the academy also. It hit me most fiercely when thinking about what is going on in Afghanistan, and in the Swat valley, young girls getting acid thrown over them when they try to go to school, and what our commitment should be. How do feminist academics, feminists in education, take this feminist, educational issue seriously, and how do we act upon our address to the fact of young women being denied education? I think that poses a lot of tricky issues for us in relation to a crosscutting of our own values, political positions, with multicultural sensitivity, or intercultural sensitivity. I think we have to address that crosscutting a lot more. We do the cultural sensitivity bit really a lot these days, but standing up for other things in those contexts can be extremely tricky. I'd like to have a context in which to think about something like that. So I think there are issues beyond the academy also, but within education if you like, that we might be thinking about.

Could you elaborate on this insistence that doing geography as a feminist is about much more than foregrounding gender? How does this rework understandings of geography?
First, then, the point is the feminism of my generation wasn't just about gender, it was about much more general liberation. It was, to some extent, a response to the '60s, which were very male dominated, but it was part of that real challenge to hierarchies and all the rest of it. It was a question of style and democracy and therefore I'd want that to be part of being a feminist in the academy.

Such changes have been brought by feminism into the academy. Nonetheless, I think there's a real irony in the present moment of so-called postmodernism, post-structuralism or whatever, about big global stars. I do really dislike it. I hate that kind of strutting of the world's stage and there is a certain pomposity that can come with the 'big name' structuring of the academy (I'm thinking here mainly of philosophy and critical theory, not geography). And also a kind of genuflection by others; people seem to think there are big names they have to read and they have to decide which one they're ... which team they're in, almost, you know. Some people, including people who would call themselves feminists, are still perhaps too in awe of some of the great figures and have a concept of a great figure, which I find ... I just don't like that much.

And there's little things associated with that, like I've never had or done blurbs on the back of books and that partly came out of that. I know it is highly debatable, and I know there are good sides to it, but it's all about name, it's all about validation by big names and stuff like that.

Second, there's a whole set of issues around the genealogy and politics of some of the conceptual advances of recent decades. The degree to which that conceptual questioning emerged from and was developed within polit-ical struggle – including, very importantly, feminist struggles – seems to me to be under-recognised. Many of the explorations of the construction of identity, all the arguments against essentialism, perhaps relational thinking more generally, were importantly first explored in debates within political constituencies – anti-racism, feminism and sexual liberation. It is a great pity when those roots are forgotten; conceptual debate can then lose its urgency and real meaning. That real two-way movement between the con-ceptual and the political has always, for me, been very important.

And third, and more generally, there is a real issue, at least for me, about how we 'live' our theoretical positions. I don't mean this in a pious way at all (though the glaring disjunction sometimes between the criticisms of competitive individualism that occur so often in our writings and the actual competitive individualism within the academy is dismaying). But there are other, funnier, examples. There all those people who pen 'daring' theory about dislocating this and fracturing the other … and so forth, and then go home and live the most boring and conventional, and unchallenging, lives. Or there are those people who have written most about the death of the author but who are often most into the kind of personality culture within the academy. So that … and I do think that's not feminist in the old sense of 'feminism'. Likewise, and this is a bit of an aside, but all those people that go on about everything being tentative, hesitant, fleeting, precarious – they're often the ones that want every word recorded for somewhere on their websites, you know. It's a real kind of desire to fix. I was just in a conference and somebody who is a real Deleuzean came up … couldn't be at the session … very sorry, could we tape it or whatever? And I said, no, it was fleeting and you missed it. Again, it's about how (or whether) we 'live' our theoretical positions.

And there's another (fourth) thing, that I don't know how to put, because I don't want to be essentialist. This is a political position, not an essentialisation around masculinity, femininity or whatever. But I do find myself amazed by and wary of the ease with which writers make Olympian statements about the age and the changes in the age we are living in. And of course, that is what gets picked up by journalists and popular press and so forth. I absolutely don't want to associate feminism with the local. I'm even more opposed to that! I think feminists ought to do global stuff too, so it's not that. It's for other reasons: a kind of commitment to complexity which, thinking back to even *Spatial Divisions of Labour*, has always been there. That kind of commitment to complexity but also a feeling that maybe we can *change* the way things are. Those who are making these grand statements are often slightly standing outside society and describing it and forgetting that we're also within it. The notion of being *within* the society slightly changes one's ability to make Olympian statements. So I'm not sure about all of this, but it's somehow bound up with having been in that second-wave feminist movement.

I think this is both class and feminism, but I don't know … I have a very strong sense of not being part of the establishment. I've turned down a number of establishment offerings because they just didn't fit. It's just … you know one has got lots of prizes and all the rest of it, one is reasonably well known [laughs] but I still don't feel … and I think that's a mixture of class and feminism. And that affects what kind of voice you have; what kind of role you can play, and want to play.

Could you tell us a bit about your recent work in Venezuela and, specifically, how the concept of power-geometries has been taken up by Chávez and the Bolivarian revolution?
Well, the work on space and power led to the emergence of the notion of power-geometries. By that I was trying to capture at the same time the fact

that space is always a product of power-filled social relations and the fact that power always has a geography.

The idea of power-geometries was picked up by Hugo Chávez through some sociologists and geographers.[11] He reads enormously, he is quite astonishing. He will sit there on *Alo Presidente*, which is a television programme which goes on for endless hours on Sunday; like watching a test match, you can come in and out of it, it goes on and on. Each one is held in a different place – a cooperative or a new project, for instance. And Chávez will be sitting there talking and he will pick up a book and read a paragraph and put it down and he will say, 'What I think that is trying to tell us is ...' There is a real popular pedagogy going on there that I find extremely interesting and there is no doubt that he reads, and thinks, extremely widely and openly. And somebody had told him about power-geometries and given him, I guess, edited versions of some of the stuff that I'd written. And he picked up on it and it became, of the five motors of the revolution which were inaugurated in 2007, number 4, which is 'we have to build a new power-geometry'.

Here in Venezuela the concept of power-geometry is being used primarily to think about the political reorganisation of the country. It is immensely impressive. If you're going to have a new society you need also to build a new geography, especially a new geography of power, of political voice, and the span of this intervention is wide, from addressing the inequality between regions and within cities, to the active enablement of the growth of grass-roots power.

This is partly by an equalisation of voice between the southern regions and the coastal regions. So there is a huge emphasis on trying to get democratic structures going in the more rural and southern parts of the country, to give them more voice. That, I think, is a long and difficult process. There is no doubt that the coastal regions dominate. But at least now there is something on the agenda which makes it a question. There is also, across the country, including the cities, the establishment of a structure of participatory, direct democracy parallel to the state. So every self-defined group of 400 or so households has the right to set up a communal council, Consejo Comunal, and that communal council has to conform to certain things, put forward plans, all the rest of it. But then it has rights and resources to self-government. The long-term idea is that within the local area there can be self-government and it can be a voice to put pressure on the parallel structures of the elected state, which before Chávez had fallen into such disarray. And those communal councils are then engaged in a process of aggregation. It's a way of building a political structure of the country from the base upwards.

Remember also, this is not brand new in Latin America. In Chile under Allende there were locally based councils as well; it does have quite a long history. But I do think it is very interesting and I have spent a bit of time talking to the people who are trying to form and organise these Consejos Comunales. They are often women, often older women, and there is one conversation that struck me very much: a woman who was organising a council in Petare, which is an area outside Caracas but adjacent to it – a vast area of informal housing, we would call it. And she said to me how wonderful the idea of communal councils was; it was brilliant. She talked about all the problems within the area. There is an assumption in this initiative that there is some kind of coherence to place and that people will agree, and of course what has happened is that there are battles happening within these household groupings. On the one hand that is quite destructive, but on the other hand it is a process of political education. But the thing she said to me which really made me think was, 'Que voy a hacer con todo este poder?' ('What am I going to do with all this power?') And it was like the opposite of what we always say in geography – you know, give 'local people' the power and all the rest of it – and she said, 'antes de darnos' ('before you give it to us'), 'hay que enseñarnos' ('you have to teach us'). In other words, the very process of setting up these things is a process of learning and it is going to be very difficult and it is going to take a lot of time.

And I have learned a lot. My appreciation of the forms and nuances of power was vastly enriched; the concept of power-geometries itself was dynamised – precisely because it was being used in a political *process*; I have been provoked to think yet more about the potential relations between places and politics. So far, everything I've written about it has been in Spanish, as part of the process, but I do hope to reflect on it more in English (there is much we can learn from Latin America) – the problem, as always, is finding the time![12]

One thing that's been taking up your time recently has been working on a 'landscape' project with the film-maker Patrick Keiller[13]
Yes. It might be worth saying something about working with people in 'the arts', since I've done so on quite a few occasions in lots of different ways. I'm wary of the vanity-catalogue phenomenon, but I have found some really constructive ways to intersect. Many artists think a lot about space, and they think about it very differently, so good conversations can dislodge one from the customary tramlines.

For example, I've worked quite a bit with Olafur Eliasson, especially on his *Weather project* at Tate Modern where I both wrote with/for him,[14] as I have done on a number of occasions (see his website), and organised with him a great seminar series at Tate Modern. (In fact, I did a piece in the original catalogue when Tate Modern opened.)[15]

Anyway, this project with Patrick Keiller (and also Patrick Wright) had the additional attractions that it is a properly joint project whose central concern is with landscape.

How have you been thinking landscapes politically in relation to this?
You are right to pinpoint that. A central thread of the project is concerned
with reading landscape politically. We meant this in the sense of asking what
one can diagnose of the current political situation (in its longer historical
context) through an exploration of (and, I would add, a particular conception of) landscape. For Patrick this also involved a specific approach – that
of using film-making as a research method. For me, it involved bringing
together my work on space, the form and process of film-making, and the
landscape itself. So I've written a long essay that is a kind of companion
piece to the film. There is now a DVD available containing both the film
(*Robinson in Ruins*), directed by Patrick, and my essay ('Landscape/space/
politics').[16]
'Thinking landscape politically' in this context had three strands to it:
about the form of the film, about the stories in the landscape, and about
specific political themes which emerged in the process of our exploration.

The first of these – the form of the film – has been for me particularly
interesting because, it turns out, the way Patrick has constructed the film
resonates with the way in which I have for ages been arguing we should
conceptualise space. On his journey in the landscape he only films when he
stops, and the camera is still. He moves, stops, focuses on something, and
films. And each time he stops there's a story to be drawn out of the landscape.
He's not walking across space as a surface but giving due to a multiplicity of
stories. This is landscape as a multiplicity of stories-so-far, which is exactly
how I have conceptualised space. That is exactly … that formal structure of
the film mirrors what I've been arguing about space. It was only deep into
the project that I clicked why I responded so much to what he was doing. It
was a moment of real 'conceptual happiness', if you know what I mean.

Now, my argument would be that that opens the landscape up politically
in a way that other readings and political critiques of landscape don't necessarily do. It avoids, for instance, the over-easy recourse to notions of palimp-
sest. It is more challenging than many 'heritage' critiques. Both of these
tend to be critiques of the past that leave things in the past. They don't
challenge today. (The very making of the critique can make you feel better
about yourself. It doesn't necessarily challenge you.) But if you think of
space and landscape as stories-so-far, then the stories are ongoing; they
arrive at today.

And that allows you to say the contestation of that story continues.
Exactly, exactly. And for me that is crucial. My concern has long been
about contesting those imaginations of space as 'the fixed, the dead' and so
forth – to bring it alive (and the essay also argues that the film does this
through its long stillnesses that are not stasis; I argue that one can read
this contrapuntally alongside Bergson). And not only to bring space alive,
but to bring it alive *politically*. So this notion of stories-so-far stresses also

the producedness of space and the fact that, above all, space is the dimension of multiplicity. And, fully appreciated, as the film also evokes, that notion of space enjoins us to look outwards, across and around the planet, at the myriad trajectories contemporaneous with, and sometimes interdependent with, our own. In a sense this is the counterpoint to early relational arguments about place (and indeed identity more generally), where the stress was upon internal multiplicity. That is important to recognise, but a focus on that alone can become solipsistic. We should also be aware of relations and practices that spread outwards – cultivate what I call 'outwardlookingness'. From early pieces like 'In what sense a regional problem?'[17] I've been pursuing these thoughts about the conceptualisation of space – and they are brought together in *For Space*.[18]

Anyway, the second thing I've done in the essay is pull out some of the stories themselves. The film is very spare, so I've elaborated the stories, pulled them out of the landscape to pose questions for today.

And, third, this has coalesced into three themes that relate to the present conjuncture. The first is to challenge the currently hegemonic understanding that markets are natural. They are not. What's more, in their unfettered operation around the planet they destroy what we might actually call 'nature'.

The power of the non-human is strong in the film and this has long been important to me. To celebrate my 60th birthday I went on a small ex-Soviet research boat into the high Arctic (the photograph on the cover of *For Space* is one I took on that trip). It affected me a lot. Immersed in birds, and geology and the pack ice. The combination of their implacable otherness and indifference to us humans, and yet their vulnerability, as the ice melts. This was a mighty landscape, but Patrick produces the same effect (for me) with his long lingering on a foxglove – its otherness, its indifference, its vulnerability, and yet, when all is said and done perhaps, the marginality of the human.

The second argument is a critique of commodification. The film unearths, among others, stories of enclosure. These were the early moments of the commodification of both land and labour and the formation of national markets for both, with the attendant social disruption. Today we see the same processes at global level – the formation of a global labour market, for instance. Moreover, the financial crisis and the ecological crisis overarch the film and both are bound up with commodification – of risk in the first case and in the second case the doomed attempt at a solution through the commodification of carbon.

And third, and finally, we went into the project against a background of notions of 'belonging' to a landscape. For a host of reasons, though, our exploration led us to ask not 'Do we belong to this landscape?' but 'To whom does this landscape belong?' At a moment of increasing inequality and elite hegemony, this seems urgent.

All these three themes relate back to the question of the current conjuncture. I would argue that though in the global North there is an economic crisis, there has been no fracturing of the hegemonic ideology that intersects with and underpins the present economy. All these arguments, posed by the film and the essay, are meant to contribute to such a fracturing. Without a wider ideological and political questioning, there can be no conjunctural rupture nor any change in the balance of social forces.

Notes

1   Holloway (2002).
2   Massey (1991b).
3   See Hardt and Negri (2001), Holloway (2002) and Massey (2005).
4   Milne (2010).
5   Massey (1973).
6   Massey (2007).
7   Massey (2011a).
8   Massey (1991c).
9   Soja (1989) and Harvey (1989).
10 See also Deutsche (1991), Morris (1992b), and the response by Harvey (1992).
11 See Massey (1993a) and (1999a).
12 Actually, a paper is just appearing: Massey (2011c).
13 This project, 'The Future of Landscape and the Moving Image', was funded by the Arts and Humanities Research Council. The project website is at: http:// thefutureoflandscape.wordpress.com/ (last accessed 23 July 2012). Patrick Keiller is 'one of Britain's most intellectually stimulating filmmakers ... widely acclaimed for *London* (1994), his extraordinary portrait of the UK capital, and *Robinson in Space* (1997)' (British Film Institute, 2011).
14 Massey (2003).
15 Massey (2000c).
16 The DVD is available from the British Film Institute, cat. no. BFIB1098. The essay is at http://thefutureoflandscape.wordpress.com/landscapespacepolitics-an-essay/ (last accessed 23 July 2012) as well as on the DVD.
17 Massey (1979).
18 Massey (2005).

References

References to the Work of Doreen Massey

Massey, D. (1973) 'Towards a critique of industrial location theory', *Antipode* 5: 33–39.

Massey, D. (1978) 'Regionalism: some current issues', *Capital and Class* 6: 106–125.

Massey, D. (1979) 'In what sense a regional problem?', *Regional Studies* 13, 233–243.

Massey, D. (1983a) 'The contours of victory … dimensions of defeat', *Marxism Today* (July): 16–19.

Massey, D. (1983b) 'The shape of things to come', *Marxism Today* (April): 18–27.

Massey, D. (1983c) 'Industrial restructuring as class restructuring: production decentralization and local uniqueness', *Regional Studies* 17 (2): 73–89.

Massey, D. (1984) *Spatial Divisions of Labour: Social Structures and the Geography of Production*. Basingstoke: Macmillan.

Massey, D. (1985) 'New directions in space', in D. Gregory and J. Urry (eds), *Social Relations and Spatial Structures*. London: Macmillan, pp. 9–19.

Massey, D. (1986) 'Nicaragua: some reflections on socio-spatial issues in a society in transition', *Antipode* 18 (3): 322–331.

Massey, D. (1987) *Nicaragua: Some Urban and Regional Issues in a Society in Transition*. Milton Keynes: Open University Press.

Massey, D. (1991a) 'The political place of locality studies', *Environment and Planning D: Society and Space* 23: 267–281.

Massey, D. (1991b) 'A global sense of place', *Marxism Today* (June): 24–29.

Massey, D. (1991c) 'Flexible sexism', *Environment and Planning D: Society and Space* 9: 31–57.

Massey, D. (1992) 'Politics and space/time', *New Left Review* 196: 65–84.

Spatial Politics: Essays for Doreen Massey, First Edition.
Edited by David Featherstone and Joe Painter.
© 2013 John Wiley & Sons, Ltd. Published 2013 by John Wiley & Sons, Ltd.

Massey, D. (1993a) 'Power-geometry and a progressive sense of place', in J. Bird, B. Curtis, T. Putnam, G. Robertson, and L. Tuckner (eds), *Mapping the Futures: Local Cultures, Global Chance.* London: Routledge, pp. 59–69.

Massey, D. (1993b) 'Classics in human geography revisited: author's response', *Progress in Human Geography* 17 (1): 71–72.

Massey, D. (1994a) *Space, Place and Gender.* Cambridge: Polity Press.

Massey, D. (1994b) 'Double articulation', in A. Bammer (ed.), *Displacements: Cultural Identities in Question.* Bloomington: Indiana University Press, pp. 110–122.

Massey, D. (1995a) 'Thinking radical democracy spatially', *Environment and Planning D: Society and Space* 13: 283–288.

Massey, D. (1995b) *Spatial Divisions of Labour: Social Structures and the Geography of Production* Macmillan: London.

Massey, D. (1997a) 'Editorial: problems with globalization', *Soundings* 7: 7–12.

Massey, D. (1997b) 'Spatial disruptions', in S. Golding (ed.), *The Eight Technologies of Otherness.* London: Routledge.

Massey, D. (1997c) 'A feminist critique of political economy', *City* 2 (7): 156–162.

Massey, D. (1999a) *Power-geometries and the Politics of Space-Time.* Heidelberg: Department of Geography, University of Heidelberg.

Massey, D. (1999b) 'Space-time, 'science' and the relationship between physical geography and human geography', *Transactions of the Institute of British Geographers* 24: 261–276.

Massey, D. (1999c) 'Imagining globalization: power-geometries of time-space', in A. Brah, M. J. Hickman and M. Mac an Ghaill (eds), *Global Futures: Migration, Environment and Globalization.* Basingstoke: Macmillan, pp. 27–44.

Massey, D. (1999d) 'On space and the City', in D. Massey, J. Allen and S. Pile (eds), *City Worlds.* London: Routledge, pp. 151–174.

Massey, D. (2000a) 'Travelling thoughts', in P. Gilroy, L. Grossberg and P. McRobbie (eds), *Without Guarantees: In Honour of Stuart Hall.* London: Verso, pp. 225–232.

Massey, D. (2000b) 'Space-time, 'science' and the relationship between physical geography and human geography', *Transactions of the Institute of British Geographers* 24(3): 261–276.

Massey, D. (2000c) 'Bankside: international local', in *Tate Modern: The Handbook.* London: Tate Gallery Publishing, pp. 24–27.

Massey, D. (2001) 'Talking of space-time', *Transactions of the Institute of British Geographers* 26 (2): 257–261.

Massey, D. (2002) 'Don't let's counterpose place and space', *Development* 45 (2): 24–25.

Massey, D. (2003) 'Some times of space', in S. May (ed.), *Olafur Eliasson: The Weather Project.* London: Tate Publishing, pp. 107–118.

Massey, D. (2004) 'Geographies of responsibility', *Geografiska Annaler B* 86 (1): 5–18.

Massey, D. (2005) *For Space.* London: Sage.

Massey, D. (2006) 'Landscape as a provocation: reflections on moving mountains', *Journal of Material Culture* 11 (1/2): 33–48.

Massey, D. (2007) *World City.* Cambridge: Polity.

Massey, D. (2008) 'When theory meets politics', *Antipode* 40 (3): 492–497.

Massey, D. (2010) 'The political struggle ahead', *Soundings* 45: 6–18.

Massey, D. (2011a) 'Economics and ideology in the present moment', *Soundings* 48: 29–39.

Massey, D. (2011b) 'A counterhegemonic relationality of place', in E. McCann and K. Ward (eds), *Mobile Urbanism: Cities and Policy Making in the Global Age*. Minneapolis: University of Minnesota Press, pp. 1–14.

Massey, D. (2011c) 'Espacio y sociedad: experimentos con la espacialidad del poder y democracia', in A.G. González (ed.), *Latinoamérica, laboratorio mundial*. Proceedings of IV Seminario Atlántico de Pensamiento, Gran Canaria: Madrid, La Oficina Ediciones, pp. 29–45.

Massey, D. and Catalano, A. (1978) *Capital and Land: Landownership by Capital in Great Britain*. London: Edward Arnold.

Massey, D., Human Geography Research Group, Bond, S. and Featherstone, D. (2009) 'The politics of place beyond place: a conversation with Doreen Massey', *Scottish Geographical Journal* 125: 401–420.

Massey, D. and Livingstone, K. (2007) 'The world we're in: interview with Ken Livingstone', *Soundings* 36: 11–25.

Massey, D. and McDowell, L. (1995) 'A woman's place?', in D. Massey (ed.), *Space, Place and Gender*. Cambridge: Polity Press, pp. 191–211.

Massey, D. and Meegan, R. (1982) *The Anatomy of Job Loss. The How, Why and Where of Employment Decline*. London: Methuen.

Massey, D. and Miles, N. (1984) 'Mapping out the unions', *Marxism Today* (May): 19–22.

Massey, D. and Painter, J. (1989) 'The changing geography of trade unions', in J. Mohan (ed.), *The Political Geography of Contemporary Britain*. London: Macmillan.

Massey, D., Quintas, P. and Wield, D. (1991) *High-Tech Fantasies: Science Parks in Society, Science and Space*. London: Routledge.

Massey, D. and Wainwright, H. (1985) 'Beyond the coalfields: the work of the miners' support groups', in H. Beynon (ed.), *Digging Deeper: Issues in the Miners' Strike*. London: Verso, pp. 149–168.

Allen, J. and Massey, D. (eds) (1988) *The Economy in Question: Restructuring Britain*. London: Sage.

Allen, J., Massey, D. and Cochrane, A. with Charlesworth, J., Court, G., Henry, N. and Sarre, P. (1998) *Rethinking the Region*. London: Routledge.

Amin, A., Massey, D. and Thrift, N. (2000) *Cities for the Many Not the Few*. Bristol: Policy Press.

Amin, A., Massey, D. and Thrift, N. (2003) *Decentering the Nation: A Radical Approach to Regional Inequality*. London: Catalyst.

Hall, S., Massey, D. and Rustin, M. (1995) 'Uncomfortable times', *Soundings: A Journal of Politics and Culture* 1: 5–18.

Harrison, S., Massey, D. and Richards, K. (2006) 'Complexity and emergence (another conversation)', *Area* 38 (4): 465–471.

McDowell, L. and Massey, D. (1984) 'A woman's place?', in D. Massey and J. Allen (eds), *Geography Matters! A Reader*. Cambridge: Cambridge University Press, pp. 195–215.

Other References

Agamben, G. (2002) *Homo Sacer: o poder soberano e a vida nua I.* Belo Horizonte, Brazil: Editora UFMG.

Agnew, J.A. (2009) 'Territory', in D. Gregory, R. Johnston, G. Pratt, M.J. Watts and S. Whatmore (eds), *The Dictionary of Human Geography*, 5th edn. Oxford: Wiley-Blackwell, pp. 746–747.

Agnew, J.A. and Corbridge, S. (1995) *Mastering space: Hegemony, Territory and International Political Economy.* New York, London: Routledge.

Aguiar, L.L.M. and Ryan, S. (2009) 'The geographies of the Justice for Janitors', *Geoforum* 40 (6): 949–958.

Ahumada, J.L. (1991) 'Logical types and ostensive insight', *International Journal of Psychoanalysis* 72: 683–691.

Albert, M. (2001) 'Territoriality and modernization', Workshop *The Cluster of Water, Energy and the Human Environment: Towards an Extra Territorial Concept for the Middle East*, Dead Sea Movenpick Resort and Spa, Jordan, 20–21 October.

Alcock, P., Cochrane, A. and Lee, P. (1984) 'A parable of how things might be done differently', *Critical Social Policy* 9: 69–87.

Allen, J. (2003) *Lost Geographies of Power.* London: Sage.

Allen, J. (2009) 'Three spaces of power: territory, networks, plus a topological twist in the tale of domination and authority', *Journal of Power* 2: 197–212.

Allen, J. and Cochrane, A. (2007) 'Beyond the territorial fix: regional assemblages, politics and power', *Regional Studies* 41 (9): 1161–1175.

Allen, P. and Brooke Wilson, P. (2008) 'Agrifood inequalities: globalisation and localization', *Development* 51 (4): 534–540.

Alliès, P. (1980) *L'invention du territoire.* Grenoble Presses: Universitaires de Grenoble.

Althusser, L. (1971) *On Ideology.* London: New Left Books.

Althusser, L. (2005) *For Marx*, trans. B. Brewster. London: Verso.

Althusser, L. and Balibar, E. (2009) *Reading Capital.* London: Verso.

Amin, A. (2002a) 'Spatialities of globalisation', *Environment and Planning A* 34 (3): 385–399.

Amin, A. (2002b) 'Ethnicity and the multicultural city: living with diversity', *Environment and Planning A* 34 (6): 959–980.

Amin, A. (2004) 'Regions unbound: towards a new politics of place', *Geografiska Annaler B* 86 (1): 33–44.

Amin, A. (2006) 'The good city', *Urban Studies* 43 (5/6): 1009–1023.

Amin, A. (2010) 'Neighbourly bonds', in M. Bunting, A. Lent and M. Vernon (eds), *Citizen Ethics in a Time of Crisis.* London: The Citizens Ethics Network, p. 55.

Amin, A. and Thrift, N. (2002) *Cities: Reimagining the Urban.* Cambridge: Polity Press.

Amin, A. and Thrift, N. (2005) 'What's left? Just the future', *Antipode* 37 (2): 220–238.

Amoore, L. and Langley, P. (2003) 'Ambiguities of global civil society', *Review of International Studies* 30: 89–110.

Anderson, B. (1991) [1983] *Imagined Communities: Reflections on the Origin and Spread of Nationalism.* London: Verso.

Anderson, B. and McFarlane, C. (2011) *Assemblage and Geography: Special Issue of Area* 43 (2).

Anheier, H. and Katz, H. (2005) 'Network approaches to global civil society', in H. Anheier, M. Kaldor and M. Glasius (eds), *Global Civil Society Year Book 2005/6*. London: Sage, pp. 206–221.

Anheier, H. Glasius, M. and Kaldor, M. (2001) 'Introducing global civil society', in H. Anheier, M. Kaldor and M. Glasius (eds), *Global Civil Society 2001*. Oxford: Oxford University Press, pp. 1–22.

Antonsich, M. (2009) 'On territory, the nation-state and the crisis of the hyphen', *Progress in Human Geography* 33 (60): 789–806.

Appadurai, A. (1996) *Modernity at Large: Cultural Dimensions of Globalisation*. Minneapolis: University of Minnesota Press.

Appadurai, A. (2007) 'Hope and democracy', *Public Culture* 19 (1): 29–34.

Arendt, H. (1979) *The Origins of Totalitarianism*. London, New York: Harvest/HBJ.

Awatere, D. (1984) *Māori Sovereignty*. Auckland: Broadsheet.

Ayers, A. J. (2006) 'Beyond the imperial narrative: African historiography revisited', in B. Gruffydd Jones (ed.), *Decolonising International Relations*. London: Rowman and Littlefield, pp. 155–177.

Badie, B. (1995) *La Fin des territoires*. Paris: Fayard.

Baker, A. (2003) *Geography and History: Bridging the Divide*. Cambridge: Cambridge University Press.

Balibar, E. (2007) *The Philosophy of Marx*, trans. C. Turner. London: Verso.

Bandy, J. and Smith, J. (eds) (2005) *Coalitions Across Borders: Transnational Protest and the Neoliberal Order*. Lanham: Rowman and Littlefield.

Barber, B. (2004) *Fear's Empire: War, Terrorism and Democracy*. London: W.W. Norton.

Bargh, M. and Otter, J. (2009) 'Progressive spaces of neoliberalism in Aotearoa: a genealogy and critique', *Asia Pacific Viewpoint* 50 (2): 154–165.

Barker, K. (2004) *Review of Housing Policy. Delivering Stability: Securing our Future Housing Needs. Final Report*. Norwich: HMSO.

Barnes, T., Peck, J., Sheppard, E. and Tickell, A. (2007) 'Methods matter: transformations in economic geography', in A. Tickell, E. Sheppard, J. Peck and T. Barnes (eds), *Politics and Practices in Economic Geography*. London: Sage, pp. 1–24.

Barnett, C. and Low, M. (2004) 'Geography and democracy: an introduction', in C. Barnett, and M. Low (eds), *Spaces of Democracy: Geographical Perspectives on Citizenship, Participation and Representation*. London: Sage, pp. 1–22.

Baschet, J. (2005) *La Rébellion Zapatiste*. Paris: Champs Flammarion.

Bateson, G. (1973) *Steps to an Ecology of Mind*. London: Granada.

Beckett, F. and Hencke, D. (2009) *Marching to the Fault Line: The Miners' Strike and the Battle for Industrial Britain*. London: Constable and Robinson.

Berberoglu, B. (2010) *Globalization in the 21st Century: Labor, Capital, and the State on a World Scale*. New York: Palgrave Macmillan.

Bergene, A.C., Endresen, S.B. and Knutsen, H.M. (eds) (2010) *Missing Links in Labour Geography*. Aldershot: Ashgate.

Bergson, H. (1993) [1927] *Essai sur les données immédiates de la conscience*. Paris: Presses Universitaires de France.

Bergson, H. (2008) *Time and Free Will: An Essay on the Immediate Data of Consciousness*, trans. F.L. Pogson. New York: Cosimo.

Berlinguer, M. (2010) 'Free culture movement', *Red Pepper* (October/November).

Beynon, H. (1973) *Working for Ford*. Harmondsworth: Penguin.

Biggs, M. (1999) 'Putting the state on the map: cartography, territory, and European state formation', *Comparative Studies in Society and History* 41 (2): 374–405.

Bingham, N. (2006) 'Bees, butterflies, and bacteria: biotechnology and the politics of nonhuman friendship', *Environment and Planning A* 38 (3): 483–498.

Blanco, M. (1975) *The Unconscious as Infinite Sets*. London: Karmac.

Boltanski, L. and Chiapello, E. (2005) *The New Spirit of Capitalism*, trans G. Elliott. London: Verso.

Bond, S. (2011) 'Being in myth and community: resistance, lived existence, and democracy in a North England mill town', *Environment and Planning D: Society and Space* 29 (4): 780–802.

Bourdieu, P. (1989) 'Social space and symbolic power', *Sociological Theory* 7 (1): 14–25.

Bowlby, S.R., Foord, J. and McDowell, L. (1986) 'The place of gender in locality studies', *Area* 18 (4): 327–331.

Boyd, R. (2006) 'The value of civility', *Urban Studies* 43: 863–878.

Braun, B. (2008) 'Environmental issues: inventive life', *Progress in Human Geography* 32 (5): 667–679.

Brennan, B., Hoedeman, O., Kishimoto, S. and Terhorst, P. (2007) *Remaking Public Water*. Amsterdam: Transnational Institute.

Brenner, N. (1999) 'Globalisation as reterritorialisation: the rescaling of urban governance in the EU', *Urban Studies* 36 (3): 431–51.

British Film Institute (2011) 'Press release: DVD release of *Robinson in Ruins* – a film by Patrick Keiller'. At http://www.thespinningimage.co.uk/news/displaynews-item.asp?newsid=1940 (last accessed 24th July, 2012).

Burawoy, M. (2010) 'From Polanyi to Pollyanna: the false optimism of global labor studies', *Global Labour Journal* 1 (3): 301–313.

Burawoy, M., Blum, J.A., George, S., Gille, Z., Gowan, T., Haney, L., Klawiter, M., Lopez, S.H., Riain, S.O. and Thayer, M. (eds) (2000) *Global Ethnography: Forces, Connections and Imaginations in a Postmodern World*. London: University of California Press.

Byrnes, G. (2006) '"Relic of 1840" or founding document? The Treaty, the Tribunal and concepts of time', *Kōtuitui: New Zealand Journal of Social Sciences Online* 1: 1–12.

Callinicos, A. (2009) *Imperialism and Global Political Economy*. Cambridge: Polity Press.

Casey, E.S. (1997) *The Fate of Place*. Berkeley: University of California Press.

Castells, M. (1996) *The Rise of the Network Society*. Oxford: Blackwell.

Castells, M. (1997) *The Power of Identity*. Oxford: Blackwell.

Castells, M. (2009) *The Power of Identity: The Information Age – Economy, Society, and Culture Volume II*. Oxford: Wiley-Blackwell.

Castoriadis, C. (1997) *World in Fragments: Writings on Politics, Society, Psychoanalysis, and the Imagination*. Stanford, CA: Stanford University Press.

Castree, N. (2003) 'Environmental issues: relational ontologies and hybrid politics', *Progress in Human Geography* 27: 203–211.

Castree, N. (2007) 'Labour geography: a work in progress', *International Journal of Urban and Regional Research* 31 (4): 853–862.

Castree, N. (2009) 'Place: connections and boundaries in an interdependent world', in N.J. Clifford, S.L. Holloway, S.P. Price and G. Valentine (eds), *Key Concepts in Geography*, 2nd edn. London: Sage, pp. 153–172.

Castree, N., Coe, N.M., Ward, K. and Samers, M. (2004) *Spaces of Work: Global Capitalism and the Geographies of Labour*. London: Sage.

Ceceña, A. (2004) 'The subversions of historical knowledge of the struggle: Zapatistas in the 21st century', *Antipode* 36: 361–370.

Chakrabarty, D. (2009) 'The climate of history: four theses', *Critical Inquiry* 35: 197–222.

Charlesworth, J. and Cochrane, A. (1994) 'Tales of the suburbs: the local politics of growth in the South-East of England', *Urban Studies* 31 (10): 1723–1738.

Clark, G.L. (1985) 'Review of "Spatial divisions of labor: social structures and the geography of production"', *Economic Geography* 61 (3): 290–292.

Cochran, G. and Harpending, H. (2009) *The 10,000 Year Explosion. How Civilization Accelerated Human Evolution*. New York: Basic Books.

Cochrane, A. (1986) 'What's in a strategy? The London Industrial Strategy and municipal socialism', *Capital and Class* 28: 187–193.

Cochrane, A. (1987) 'What a difference the place makes: the new structuralism of locality', *Antipode* 19 (3): 354–363.

Cochrane, A. (2011) 'Post-suburbia in the context of urban containment: the case of the South East of England', in N. Phelps (ed.), *International Perspectives on Suburbanization: A Post-Suburban World?* Basingstoke: Palgrave Macmillan.

Coe, N.M., Dicken, P. and Hess, M. (2008) 'Global production networks: realizing the potential', *Journal of Economic Geography* 8 (3): 271–295.

Coe, N.M. and Jordhus-Lier, D.C. (2010a) 'Constrained agency? Re-evaluating the geographies of labour', *Progress in Human Geography* 35 (2): 211–233.

Coe, N.M. and Jordhus-Lier, D.C. (2010b) 'Re-embedding the agency of labour', in A.C. Bergene, S.B. Endresen and H.M. Knutsen (eds), *Missing Links in Labour Geography*. Aldershot: Ashgate, pp. 29–40.

Commission on Sustainable Development in the South East (2005) *Final Report*. London: Institute of Public Policy Research.

Commissioner for the Special Areas (1935) *First Report*, Cmnd 4957. London: HMSO.

Commissioner for the Special Areas (1936) *Second Report*, Cmnd 5090. London: HMSO.

Connolly, W.E. (2010) *A World of Becoming*. Durham, NC: Duke University Press.

Connor, W. (1994) *Ethnonationalism: The Quest for Understanding*. Princeton, NJ: Princeton University Press.

Cooke, P. (ed.) (1989) *Localities*. London: Unwin Hyman.

Cooke, P. (2006) 'Locality debates', in R. Kitchin and N. Thrift (eds), *The International Encyclopedia of Human Geography*. Oxford: Elsevier.

Cooley, M. (1984) *Architect or Bee? The Human/Technology Relationship*. Boston, MA: South End Press.

Crick, B. (1982) [1962] *In Defence of Politics*. London: Penguin.

Cumbers, A., Routledge, P. and Nativel, C. (2008) 'The entangled geographies of global justice networks', *Progress in Human Geography* 32 (2): 183–201.

Dartington, T. (2010) *Managing Vulnerability: The Underlying Dynamics of Systems of Care.* London: Karnac.

Davenhill, R. (ed.) (2007) *Looking into Later Life: A Psychoanalytic Approach to Depression and Dementia in Old Age.* London: Karnac.

Davies, N. and Williams, D. (2009) *Clear Red Water: Devolution and Socialist Politics.* London: Francis Boutle Publishers.

Day, R. (2005) 'From hegemony to affinity', *Cultural Studies* 18 (5): 716–748.

DeLanda, M. (2002) *Intensive Science and Virtual Philosophy.* London: Continuum.

Delaney, D. (2005) *Territory.* Malden, MA: Blackwell.

Deleuze, G. (1988) *Foucault.* Minneapolis: University of Minnesota Press.

Deleuze, G. (1991) *Bergsonism*, trans. H. Tomlinson and B. Hannerjam. New York: Zone Books.

Deleuze, G. and Guattari, F. (1987) *A Thousand Plateaus: Capitalism and Schizophrenia*, trans. B. Massumi. Minneapolis: University of Minnesota Press.

Deleuze, G. and Guattari, F. (1992) *O que é a filosofia?* Rio de Janeiro: Editora 34.

Deleuze, G. and Guattari, F. (1994) *What Is Philosophy?*, trans. H. Tomlinson and G. Burchell. New York: Columbia University Press.

dell'Agnese, E. and Squarcina, E. (eds) (2002) *Geopolitiche dei Balcani.* Milano: Unicopli.

della Porta, D. Andretta, M. Mosca, L. and Reiter, H. (2006) *Globalization From Below.* London: University of Minnesota Press.

Department of Defense (2006) *Quadrennial Defense Review Report.* www.defenselink. mil/qdr (last accessed 25 July 2012).

Department of State (1971) *American Foreign Policy: 1941–196.* Bureau of Public Affairs, Historical Office. New York: Arno Press.

Deutsche, R. (1991) 'Boys town', *Environment and Planning D: Society and Space* 9: 5–30.

Diamond, J. (1997) *Guns, Germs, and Steel: The Fates of Human Societies.* New York: Norton.

Dicken, P. (2004) 'Geographers and 'globalization': (yet) another missed boat?', *Transactions of the Institute of British Geographers* 29: 5–26.

Dicken, P. Kelly, P.F. Olds, K. and Wai-Chung Yeung, H. (2001) 'Chains and networks, territories and scales: towards a relational framework for analysing the global economy', *Global Networks* 1: 89–112.

Di Giminiani, D. (2007) 'Que es la nueva geometría del poder?', Aporrea.org: Comunicación Popular para la Construcción del Socialismo del siglo XXI. At http://www.aporrea.org/actualidad/a40153.html (last accessed 25 July 2012).

Dirlik, A. (2001) 'Place-based imagination: globalism and the politics of place', in R. Prazniak and A. Dirlik (eds), *Places and Politics in an Age of Globalization.* Lanham, MD: Rowman and Littlefield, pp. 15–51.

Dorling, D. and Thomas, B. (2004) *People and Places, A 2001 Census Atlas of the UK.* Bristol: Policy Press.

Duncan, J. and Ley, D. (1982) 'Structural Marxism and human geography: a critical assessment', *Annals of the Association of American Geographers* 72 (1): 30–59.

Dunn, J. (1993) *Western Political Theory in the Face of the Future.* Cambridge: Cambridge University Press.

Ehrenreich, B. and Hochschild, A.R. (eds) (2003) *Global Woman: Nannies, Maids and Sex Workers in the New Economy.* London: Granta.

Eisenstein, Z. (2008) 'Resexing militarism for the globe', in R. Riley, C.T. Mohanty and M.B. Pratt (eds), *Feminism and War*. London: Zed Books.

Elden, S. (2005) 'Missing the point: globalization, deterritorialization and the space of the word', *Transactions of the Institute of British Geographers* 30 (1): 8–19.

Elden, S. (2009) *Terror and Territory: The Spatial Extent of Sovereignty*. Minneapolis: University of Minnesota Press.

Elden, S. (2010) 'Land, terrain and territory', *Progress in Human Geography* 34 (6): 799–817.

Elkins, C. (2005) *Britain's Gulag*. London: Pimlico.

Elson, D. (1979) *Value: Representation of Labour in Capitalism*. London: CSE Books.

Elson, D. (2000) 'Socialised markets not market socialism', *Socialist Register* 36: 67–85.

Elson, D. (2006) '"Women's rights are human rights": campaigns and concepts', in L. Morris (ed.), *Rights: Sociological Perspectives*. London: Routledge, pp. 94–110.

England, K. and Lawson, V. (2005) 'Feminist analyses of work: rethinking the boundaries, gendering, and spatiality of work', in L. Nelson and J. Seager (eds), *A Companion to Feminist Geography*. Oxford: Blackwell, pp. 77–92.

Escobar, A. (2001) 'Culture sits in places: reflections on globalism and subaltern strategies of localization', *Political Geography* 20: 139–174.

Escobar, A. (2008) *Territories of Difference: Place, Movements, Life, Redes*. Durham, NC: Duke University Press.

Escobar, A. (2010) 'Postconstructivist political ecologies', in M. Redclift and G. Woodgate (eds), *International Handbook of Environmental Sociology*, 2nd edn. Cheltenham, UK: Elgar, pp. 91–105.

Fall, J. (2010) 'The enduring myth of natural boundaries', *Political Geography* 29 (3): 140–147.

Farinelli, F. (1985) 'Der kampf ums dasein als ein kampf um raum: teoria e misura dello spazio geografico dal settecento ai giorni nostri', in P. Pagnini (ed.), *Geografia per il principe. Teoria e misura dello spazio geografico*. Milano: Unicopli, pp. 29–60.

Farinelli, F. (1994) 'Squaring the circle, or the nature of political identity', in F. Farinelli, G. Olsson and D. Reichert (eds), *Limits of Representation*, Munich: Accedo, pp. 11–28.

Fath, B. and Patten, B.C. (1998) 'Network synergism: emergence of positive relations in ecological systems', *Ecological Modelling* 107: 127–143.

Featherstone, D. (2005) 'Towards the relational construction of militant particularisms: or why the geographies of past struggles matter for resistance to neoliberal globalisation', *Antipode* 37: 250–271.

Featherstone, D.J. (2008) *Resistance, Space and Political Identities: The Making of Counter-Global Networks*. Oxford: Wiley-Blackwell.

Featherstone, D.J. (2012) *Solidarity: Hidden Histories and Geographies of Internationalism*. London: Zed Books.

Fleras, A. and Spoonley, P. (1999) *Recalling Aotearoa. Indigenous Politics and Ethnic Relations in New Zealand*. Auckland: Oxford University Press.

Foley, J. (2004) *The Problems of Success. Reconciling Economic Growth and Quality of Life in the South East*. Commission on Sustainable Development in the South East, Working Paper 2. London: Institute for Public Policy Research.

Forsberg, T. (2003) 'The ground without foundation? Territory as a social construct', *Geopolitics* 8 (2): 7–24.

Foucault, M. (n.d.) 'Discourse and truth'. At http://www.foucault.info/documents/parrhesia/foucault.DT6.conclusion.en.html (last accessed 25 July 2012).

Foucault, M. (2008) *Introduction to Kant's Anthropology*, trans R. Nigro. New York: Semiotext(e).

Francis, H. (2009) *History On Our Side: Wales and the 1984/5 Miners' Strike*. Swansea: Parthian Books.

Freeman, R.B. (2006) 'People flows in globalization', *Journal of Economic Perspectives* 20 (2): 145–170.

Freeman, R.B. (2007) 'The challenge of the growing globalization of labor markets to economic and social policy', in E.A. Paus (ed.), *Global Capitalism Unbound: Winners and Losers from Offshore Outsourcing*. New York: Palgrave Macmillan, pp. 23–40.

Fukuyama, F. (1992) *The End of History and the Last Man*. London: Penguin.

Fyfe, N., Bannister, J. and Kearns, A. (2006) '(In)civility in the city', *Urban Studies* 42: 853–861.

Gallo, S. (2003) *Deleuze & a educação*. Belo Horizonte, Brazil: Autêntica.

Geras, N. (1987) 'Post-Marxism?' *New Left Review I* (163): 40–82.

Gibson-Graham, J.K. (1996) *The End of Capitalism (As We Knew It): A Feminist Critique of Political Economy*. Minneapolis: University of Minnesota Press.

Gibson-Graham, J.K. (2002) 'Beyond the global versus local: economic politics outside the binary frame', in A. Herod and M. Wright (eds), *Geographies of Power: Placing Scale*. Oxford: Blackwell, pp. 25–60.

Gibson-Graham, J.K. (2004) 'Area studies after post-structuralism', *Environment and Planning A* 36: 405–419.

Gibson-Graham, J.K. (2006) *A Post-Capitalist Politics*. Minneapolis: University of Minnesota Press.

Gibson-Graham, J.K. (2008) 'Diverse economies: performative practices for "other world"', *Progress in Human Geography* 32 (5): 613–632.

Giddens, A. (1990) *The Consequences of Modernity*. Cambridge: Polity Press.

Giddens, A. (1994) *Beyond Left and Right: The Future of Radical Politics*. Cambridge: Polity Press.

Glasman, M. (2011) 'Labour as a radical tradition', in M. Glasman, J. Rutherford, M. Stears and S. White, *The Labour Tradition and the Politics of Paradox*. London: The Oxford London Seminars/Soundings, pp. 14–34.

Glassman, J. (2001) 'From Seattle (and Ubon) to Bangkok: the scales of resistance to corporate globalization', *Environment and Planning D: Society and Space* 20: 513–533.

Gordon, I. (2003) 'Three into one: joining up the Greater South East', *Town and Country Planning* 72: 342–343.

Gordon, I., Travers, T. and Whitehead, C. (2008) *London's Place in the UK Economy 2008–09*. London: Corporation of London.

Gottmann, J. (1951) 'Geography and international relations', *World Politics* 3 (2): 153–173.

Gottmann, J. (1952) 'The political partitioning of our world: an attempt at analysis', *World Politics* 4 (4): 512–519.

Gottmann, J. (1973) *The Significance of Territory*. Charlottesville: University Press of Virginia.

Graham, J. (1998) 'Review of *Spatial Divisions of Labour: Social Structures and the Geography of Production*', *Environment and Planning A* 30 (5): 942–943.

Gramsci, A, (1971) *Selections From Prison Notebooks*. London: Lawrence and Wishart.

Grandin, G. (2004) *The Last Colonial Massacre*. London: University of Chicago Press.

Gregory, D. (1989) 'Areal differentiation and postmodern *human* geography', in D. Gregory and R. Walford (eds), *Horizons in Human Geography*. London: Macmillan, pp. 67–96.

Grossberg, L. (1997) 'Cultural studies, modern logics, and theories of globalization', in A. McRobbie (ed.), *Back to 'Reality': The Social Experience of Cultural Studies*. Manchester: Manchester University Press.

Grossberg, L. (1999) 'Speculations and articulations of globalization', *Polygraph* 11: 11–48.

Grossberg, L. (2010) *Cultural Studies in the Future Tense*. Durham, NC: Duke University Press.

Gruffydd Jones, B. (2006) 'Introduction: international relations, Eurocentrism and imperialism', in B. Gruffydd Jones (ed.), *Decolonizing International Relations*. London: Rowman and Littlefield Publishers, pp. 1–19.

Gurnah, A. (1997) *Admiring Silence*. London: Penguin.

Guthman, J. (2004) *Agrarian Dreams: The Paradox of Organic Farming in California*. Berkeley: University of California Press.

Hacking, I. (2004) *Historical Ontology*. Cambridge, MA: Harvard University Press.

Haesbaert, R. (2004) *O mito da desterritorialização*. Rio de Janeiro: Bertrand Brasil.

Haesbaert, R. (2008) 'Dilemas de conceitos: espaço-território e contenção territorial', in M. Saquet and E. Sposito (eds), *Territórios e territorialidades: teorias, processos e conflitos*. São Paulo: Expressão Popular.

Haesbaert, R. (2011) 'Cultural hybridism, identitary "anthropophagy" and transterritoriality', in J. Agnew, P. Claval and Z. Roca (eds), *Landscape, Identities and Development*. Farnham, UK, Burlington, VT: Ashgate.

Häkli, J. (2001) 'In the territory of knowledge: state-centred discourses and the construction of society', *Progress in Human Geography* 25: 403–422.

Hale, A. and Wills, J. (eds) (2005) *Threads of Labour: Garment Industry Supply Chains from the Workers' Perspective*. Oxford: Blackwell.

Hall, P., Thomas, R., Gracey, H. and Drewett, R. (1973) *The Containment of Urban England: The Planning System. Objectives, Operations, Impacts*. London: Allen and Unwin.

Hall, S. (1980) 'Race, articulation and societies structured in dominance', in UNESCO (ed.), *Sociological Theories: Race and Colonialism*. UNESCO: Paris, pp. 305–345.

Hall, S. (1991) 'The local and the global: globalization and ethnicity', in A.D. King (ed.), *Culture, Globalization and the World-System*. London: Macmillan, pp. 19–40.

Hall, S. (2011) 'The neoliberal revolution', *Soundings* 48: 9–27.

Hallward, P. (2009) 'The will of the people: notes towards a dialectical voluntarism', *Radical Philosophy* 155: 17–29.

Hannington, W. (1937) *The Problem of the Distressed Areas – An Examination of Poverty and Unemployment*. London: Victor Gollancz.

Harcourt, W. (2009) *Body Politics in Development: Critical Debates in Gender and Development*. London: Zed Books.

Harcourt, W. and Escobar. A. (2002) 'Women and the politics of place', *Development* 45 (2): 7–14.

Harcourt, W. and Escobar, A. (eds) (2005) *Women and the Politics of Place*. Bloomfield, CT: Kumarian Press.

Hardt, M. and Negri, A. (1994) *Labor of Dionysus: A Critique of the State-Form*. Minneapolis: University of Minnesota Press.

Hardt, M. and Negri, A. (2001) *Empire*. Cambridge, MA: Harvard University Press; *Império*. Rio de Janeiro/São Paulo: Record.

Hardt, M. and Negri, A. (2004) *Multitude: War and Democracy in the Age of Empire*. London: Penguin.

Hardt, M. and Negri, A. (2009) *Commonwealth*. New York: Harvard University Press.

Harvey, D. [1982] (2007) *Limits to Capital*. London: Verso.

Harvey, D. (1985) *Urbanization of Capital: Studies in the History and Theory of Capitalist Urbanization*. Baltimore, MD: Johns Hopkins University Press.

Harvey, D. (1987) 'Three myths in search of a reality in urban studies', *Environment and Planning D: Society and Space* 5 (4): 367–376.

Harvey, D. (1989) *The Condition of Postmodernity: An Enquiry into the Origins of Cultural Change*. Oxford: Blackwell.

Harvey, D. (1992) 'Postmodern morality plays', *Antipode* 24 (4): 300–326.

Harvey, D. (1996) *Justice, Nature and the Geography of Difference*. Oxford: Blackwell.

Harvey, D. (2003) *The New Imperialism*. Oxford: Oxford University Press.

Harvey, D. (2006a) *Justice, Nature and the Geography of Difference*. Oxford: Blackwell.

Harvey, D. (2006b) *A Brief History of Neo-liberalism*. Oxford: Oxford University Press.

Heidegger, M. (1962) *Being and Time*. New York: Harper and Row.

Heidegger, M. (1982) *The Question Concerning Technology and Other Essays*. New York: Harper.

Herod, A. (1997) 'From a geography of labor to a labor geography', *Antipode* 29 (1): 1–31.

Herod, A. (ed.) (1998) *Organizing the Landscape: Geographical Perspectives on Labor Unionism*. Minneapolis: University of Minnesota Press.

Herod, A. (2001) *Labour Geographies: Workers and the Landscapes of Capitalism*. London: Guildford.

Herod, A. (2010) 'Labour geography: where have we been? Where should we go?', in V. Bergene, S.B. Endresen and H.M. Knutsen (eds), *Missing Links in Labour Geography*. Aldershot: Ashgate, pp. 15–28.

Herod, A., Peck, J. and Wills, J. (2003) 'Geography and industrial relations', in P. Ackers and A. Wilkinson (eds), *Understanding Work and Employment: Industrial Relations in Transition*. Oxford: Oxford University Press, pp. 176–192.

Hiernaux, D. and Lindon, A. (1993) 'El concepto de espacio y el análisis regional', *Revista Secuencia: Revista de historia y ciencias sociales Nueva Epoca* 25: 89–110.

Hill, R. (2010) 'Fitting multiculturalism into biculturalism: Maori–Pasifika relations in New Zealand from the 1960s', *Ethnohistory* 57 (2): 291–319.

Hinchliffe, S. (2007) *Geographies of Nature: Societies, Environments, Ecologies*. London: Sage.

Hobsbawm, E. (1978) 'The forward march of labour halted?', *Marxism Today* (September): 279–286.

Hobson, J.A. (1988) [1902] *Imperialism – A Study*. London: Unwin Hyman.

Holland, E. (1996) 'Schizoanalysis and Baudelaire: some illustrations of decoding at work', in P. Patton (ed.), *Deleuze: A Critical Reader*. Oxford: Blackwell.

Holland, S. (1975) *The Socialist Challenge*. London: Quartet Books.

Holloway, J. (2002) *Change the World Without Taking Power*. London: Pluto Press.

Holloway, S.L., Rice, S P. and Valentine, G. (eds) (2003) *Key Concepts in Geography*. London: Sage.

Huijser, H. (2005) 'Negotiating multicultural difference in a bicultural nation: a focused case study', in T. Khoo (ed.), *The Body Politic: Racialised Political Cultures in Australia*. Refereed Proceedings from the UQ Australian Studies Centre Conference, Brisbane, 24–26 November 2004. http://henkhuijser.webs.com/researchpublications.htm (last accessed 25 July 2012).

Humpage, L. and Fleras, A. (2001) 'Intersecting discourses: closing the gaps, social justice and the Treaty of Waitangi', *Journal of Social Policy New Zealand* 16: 37–53.

Hutton, G. (2005) *Coal Not Dole: Memories of the 1984/85 Miners' Strike*. Catrine, UK: Stenlake Publishing.

Hylton, F. and Thomson, S. (2007) *Revolutionary Horizons*. London: Verso.

IMF [International Monetary Fund] (2007) *World Economic Outlook*. Washington, DC: International Monetary Fund.

Jackson, R. (2007) *Sovereignty: Evolution of an Idea*. Cambridge: Polity Press.

James, I. (2005) 'On interrupted myth', *Journal for Cultural Research* 9 (4): 331–349.

Jamoul, L. and Wills, J. (2008) 'Faith in politics', *Urban Studies* 45 (10): 2035–2056.

Johnson, C. (2004) *The Sorrows of Empire*. London: Verso.

Johnson, J.T. (2008) 'Indigeneity's challenges to the white settler-state: creating a third space for dynamic citizenship', *Alternatives: Global, Local, Political* 33 (1): 29–52.

Johnston, R. (2001) 'Out of the "moribund backwater": territory and territoriality in political geography', *Political Geography* 20 (16): 677–693.

Jones, M. (2009) 'Phase space: geography, relational thinking, and beyond', *Progress in Human Geography* 33 (4): 487–506.

Judt, T. (2007) [2005] *Postwar: A History of Europe since 1945*. London: Pimlico.

Judt, T. (2010) *Ill Fares the Land. A Treatise on Our Present Discontents*. London: Allen Lane.

Kaldor, M. (2003) *Global Civil Society*. Cambridge: Polity Press.

Kāwharu, I.H. (ed.) (1989) *Waitangi. Maori and Pakeha Perspectives of the Treaty of Waitangi*. Oxford: Oxford University Press.

Keane, J. (2003) *Global Civil Society*. Cambridge: Cambridge University Press.

Keane, J. (2009) *The Life and Death of Democracy*. London: Pocket Books.

Kemmis, D. (1990) *Community and the Politics of Place*. Norman: University of Oklahoma Press.

Klare, M. (2002) *Resource Wars*. New York: Henry Holt/Owl Books.

Klauser, F.R. (n.d.) *Rethinking the Relationships between Society and Space: A Review of Claude Raffestin's Conceptualisation of Human Territoriality*. Working Paper.

Kramer, P.A. (2006) 'Race-making and colonial violence in the US empire: the Philippine–American war as race war', *Diplomatic History* 30 (2): 169–210.

Laclau, E. (1990) *New Reflections on the Revolutions of Our Time*. London: Verso.

Laclau, E. (2001) 'Democracy and the question of power', *Constellations* 8 (1): 3–14.

Laclau, E. (2005) *On Populist Reason*. London: Verso.

Laclau, E. and Mouffe, C. (2001) [1985] *Hegemony and Socialist Strategy: Towards a Radical Democratic Politics*, 2nd edn. London: Verso.

Larner, W. (2002) 'Neoliberalism and *Tino Rangatiratanga*: welfare state restructuring in Aotearoa/New Zealand', in C. Kingfisher (ed.), *Western Welfare in Decline. Globalisation and Women's Poverty*. Philadelphia: University of Pennsylvania Press, pp. 147–163.

Larner, W. and Craig, D. (2005) 'After neoliberalism? Community activism and local partnerships in Aotearoa New Zealand', *Antipode* 37 (3): 402–424.

Latour, B. (1993) *We Have Never Been Modern*. Hemel Hempstead, UK: Harvester Wheatsheaf.

Latour, B. (2004) *Politics of Nature: How to Bring the Sciences into Democracy*. Cambridge, MA: Harvard University Press.

Latour, B. (2005) *Reassembling the Social: An Introduction to Actor-Network-Theory*. Oxford: Oxford University Press.

Lebow, R.N. (2010) *A Cultural Theory of International Relations*. Cambridge: Cambridge University Press.

Lee, R. and Wills, J. (eds) (1997) *Geographies of Economies*. London: Arnold.

Lefebvre, H. (1976) 'Reflections on the politics of space', *Antipode* 8 (2): 30–37.

Lefebvre, H. (1991) *The Production of Space*, trans. D. Nicholson-Smith. Oxford: Blackwell.

Lefebvre, H. (2006) 'Beyond structuralism', in S. Elden, E. Lebas and E. Kopfman (eds), *Basic Writings*. London: Athlone, pp. 21–25.

Le Heron, R. and Lewis, N. (2007) 'Globalising economic geographies in the context of globalising higher education', *Journal of Geography in Higher Education* 31 (1): 5–12.

Lester, T. (2009) *The Fourth Part of the World*. New York: Free Press.

Leunig, T., Swaffield, J. and Hartwich, O. (eds) (2008) *Cities Unlimited. Making Urban Regeneration Work*. London: Policy Exchange.

Li, T. (2007) *The Will to Improve. Governmentality, Development, and the Practice of Politics*. Durham, NC: Duke University Press.

Lier, D.C. (2007) 'Places of work, scales of organising: a review of labour geography', *Geography Compass* 1 (4): 814–833.

London Edinburgh Weekend Return (1979) *In and Against the State*. London: Pluto Press.

Lopez, S.H. (2004) *Reorganizing the Rust Belt: An Inside Study of the American Labor Movement*. Berkeley: University of California Press.

Lovering, J. (1989) 'The restructuring debate', in R. Peet and N. Thrift (eds), *New Models in Geography*, Vol. 1. London: Unwin Hyman, pp. 198–223.

Lovering, J. (1990) 'Neither fundamentalism nor "new realism": a critical realist perspective on current divisions in socialist theory', *Capital and Class* 42: 30–54.

Mabey, R. (1996) *Flora Britannica*. London: Sinclair-Stevenson.

Malpas, J.E. (1999) *Place and Experience: a Philosophical Topography*. Cambridge: Cambridge University Press.

Malpas, J. (2007) *Heidegger's Topology: Being, Place, World*. Cambridge, MA: MIT Press.

Markoff, J. (1999) 'Where and when was democracy invented?', *Studies in Comparative Society and History* 41 (4): 660–690.

Marsh, D. (1989) 'British trade unions in a cold climate', *West European Politics* 12 (4), 192–198.

Marston, S.A., Jones III, J.P. and Woodward, K. (2005) 'Human geography without scale', *Transactions of the Institute of British Geographers* 30 (4): 416–432.

Martin, I. (2001) 'Dawn of the living wage: the diffusion of a redistributive municipal policy', *Urban Affairs Review* 36 (4): 470–496.

Martin, R., Sunley, P. and Wills, J. (1996) *Union Retreat and the Regions:The Shrinking Landscape of Organised Labour*. London: Jessica Kingsley.

Marvin, S., Harding, A. and Robson, B. (2006) *A Framework for City-Regions*. Working Paper 4: *The Role of City-Regions in Regional Economic Development Policy*. London: Office of the Deputy Prime Minister.

McDougall, J. (1986) *Theatres of the Mind: Illusion and Truth on the Psychoanalytic Stage*. London: Free Association Books.

McDowell, L. (2008) 'Thinking through work: complex inequalities, constructions of difference and trans-national migrants', *Progress in Human Geography* 32 (4): 491–507.

McFarlane, C. (2011) 'The city as assemblage: dwelling and urban space', *Environment and Planning D: Society and Space* 29: 649–671.

McGrath-Champ, S., Herod, A. and Rainnie, A. (eds) (2010) *Handbook of Employment and Society:Working Space*. Cheltenham, UK: Edward Elgar.

McGuinness, M. (2000) 'Whiteness and contemporary geography', *Area* 32 (2): 225–230.

McKay, B. and Walmsley, A. (2003) 'Māori time: notions of space, time and building form in the South Pacific', *IDEA Journal* 85–95. At http://www.idea-edu.com/Journal/2003/Maori-Time-Notions-of-Space-Time-and-Building-Form-in-the-South-Pacific (last accessed 25 July 2012).

MacKinnon, D. (2011) 'Reconstructing scale: towards a new scalar politics', *Progress in Human Geography* 35 (1): 21–36.

Mackinnon, D., Cumbers, A., Featherstone, D., Ince, A. and Strauss, K. (2011) *Globalisation, Labour Markets and Communities in Contemporary Britain*. York: Joseph Rowntree Foundation.

Mackintosh, M. and Wainwright, H. (1987) *A Taste of Power: The Politics of Local Economics*. London: Verso.

McNeill, J.R. and McNeill, W.H. (2003) *The Human Web: A Bird's Eye View of World History*. NewYork: Norton.

Melucci, A. (1996) *Challenging Codes*. London: Cambridge University Press.

Menéndez, J. (1991) *Arquitecturar es hacer arquitectura*. Cota: TE.

Menéndez, R. (2007) *Los Modelos de Localización de Actividades Económicas a la Luz del Concepto de espaciotiempo geográfico: El caso específico de las áreas marginales de Caracas*. Caracas: Editorial Fundación para la Cultura Urbana.

Mertes, T. (2002) 'Grass-roots globalism', *New Left Review* 17: 101–110.

Mertes, T. (ed.) (2004) *A Movement of Movements: Is Another World Really Possible?* London: Verso.

Milkman, R. (2000) 'Immigrant organizing and the New Labor Movement in Los Angeles', *Critical Sociology* 26 (1–2): 59–81.

Miller, D. (2011) 'Value, abstraction and material'. At http://spatialdisjunctures. wordpress.com/2011/10/26/value-abstraction-and-material/(last accessed 25 July 2012).

Milne, S. (2010) 'A people power fraud that promises mass privatisation'. *The Guardian*, Thursday 15 April: http://www.guardian.co.uk/commentisfree/2010/apr/15/big-society-conservatives-people-power?INTCMP=SRCH (last accessed 25 July 2012).

Ministry of Labour (1934) *Report of the Inquiry into the Industrial Conditions in Certain Depressed Areas*, Cmnd 4728. London: HMSO.

Mitchell, D. (2005) 'Working-class geographies: capital, space, and place', in J. Russo and S.L. Linkon (eds), *New Working-class Studies*. Ithaca, NY: Cornell University Press, pp. 78–97.

Mitchell, H. (1977) *The Hard Way Up: The Autobiography of Hannah Mitchell, Suffragette and Rebel.* London: Virago.

Mitchell, S. and Aron, L. (1999) *Relational Psychoanalysis: The Emergence of a Tradition.* London: Routledge.

Money-Kyrle, R. (1971) 'The aim of psychoanalysis', *International Journal of Psychoanalysis* 53: 103–106.

Moody, K. (2004) 'Workers of the world', *New Left Review* 27: 153–160.

Moore, B. and Rhodes, J. (1973) 'Evaluating the effects of British regional economic policy', *The Economic Journal* 83: 87–110.

Morgan, K. (2007) 'The polycentric state: new spaces of empowerment and engagement?', *Regional Studies* 41 (9): 1237–1251.

Morris, M. (1992a) 'On the beach', in L. Grossberg, C. Nelson and P. Treichler (eds), *Cultural Studies*. New York: Routledge.

Morris, M. (1992b) 'The man in the mirror: David Harvey's "condition" of postmodernity', *Theory Culture Society* 9 (1): 253–279.

Mouffe, C. (1995) 'Post Marxism, democracy and identity', *Environment and Planning D: Society and Space* 13: 259–265.

Mouffe, C. (2000) *The Democratic Paradox*. London: Verso.

Mouffe, C. (2005) *On the Political*. Abingdon: Routledge.

Murphy, A.B. (1990) 'Historical justifications for territory claims', *Annals of the Association of American Geographers* 80 (4): 531–648.

Murphy, A.B. (2002) 'National claims to territory in the modern state system: geographical considerations', *Geopolitics* 7 (2): 193–214.

Murphy, A.B. (2012) 'Territory's continuing allure', *Annals of the Association of the American Geographers*, published in early view. http://www.tandfonline.com/doi/abs/10.1080/00045608.2012.696232 (accessed 11 September 2012).

Nancy, J.-L. (1991) *The Inoperative Community*, trans. P. Connor, L. Garbus, M. Holland, and S. Sawhney. Minneapolis: University of Minnesota Press.

Nancy, J.-L. (1993) *The Experience of Freedom*. Stanford, CA: Stanford University Press.

Nancy, J.-L. (2000) *Being Singular Plural*. Stanford, CA: Stanford University Press.

Nancy, J.-L. (2007) *The Creation of the World, or Globalization*. Albany: SUNY Press.

Newman, D. (1999) 'Real spaces, symbolic spaces: interrelated notions of territory in the Arab–Israeli conflict', in P.F. Diehl (ed.), *A Road Map to War: Territorial Dimensions of International Conflict*. Nashville, TN: Vanderbilt University Press, pp. 3–34.

Newman, D. (2003) 'Boundaries', in J. Agnew, K. Mitchell and G. Toal (eds), *A Companion to Political Geography*. Malden, MA: Blackwell, pp. 123–137.

Newman, D. (2006) 'The lines that continue to separate us: borders in our "borderless" world', *Progress in Human Geography* 30 (2): 143–161.

Newman, D. and Paasi, A. (1998) 'Fences and neighbours in the postmodern world: boundary narratives in political geography', *Progress in Human Geography* 2: 186–207.

Nicholls, W. (2008) 'The urban question revisited: the importance of cities for social movements', *International Journal of Urban and Regional Research* 32: 841–859.

Nietzsche, F. (1990) [1874] *Untimely Meditations*. Cambridge: Cambridge University Press.

Norris, P. (2002) *Democratic Phoenix: Reinventing Political Activism*. Cambridge: Cambridge University Press.

November, V., Camacho-Hübner, E. and Latour, B. (2010) 'Entering a risky territory: space in the age of digital navigation', *Environment and Planning D: Society and Space* 28: 581–599.

ODPM (2003) *Sustainable Communities: Building for the Future*. London: Office of the Deputy Prime Minister.

Olesen, T. (2005) *International Zapatismo*. London: Zed Books.

Ong, A. (2007) 'Neoliberalism as a mobile technology', *Transactions, Institute of British Geographers* 32 (1): 3–8.

Ong, A. and Collier, S.J. (eds) (2005) *Global Assemblages: Technology, Politics and Ethics as Anthropological Problems*. Oxford: Blackwell.

Ong, A. (2007) 'Neoliberalism as a mobile technology', *Transactions, Institute of British Geographers* 32 (1): 3–8.

Orange, C. (2004) *An Illustrated History of The Treaty of Waitangi*. Wellington: Bridget Williams Books.

Paasi, A. (2003) 'Territory', in J. Agnew, K. Mitchell and G. Toal (eds), *A Companion to Political Geography*. Malden, MA: Blackwell, pp. 109–122.

Paasi, A. (2009) 'Bounded spaces in a "borderless world": border studies, power and the anatomy of territory', *Journal of Power* 2: 213–234.

Pain, K. (2008) 'Examining "core–periphery" relationships in a global city region: the case of London and South East England', *Regional Studies* 42: 1161–1172.

Painter, J. (2006) 'Territory-network'. Paper presented at the Annual Meeting of the Association of American Geographers, Chicago, 7–11 March. At http://dro.dur.ac.uk/8537/1/8537.pdf (last accessed 25 July 2012).

Panelli, R. and Larner, W. (2010) 'Timely partnerships? Contrasting geographies of activism in New Zealand and Australia', *Urban Studies* 47 (6): 1343–1366.

Panitch, L. (2000) 'The new imperial state', *New Left Review* 2: 5–20.

Pattie, C., Seyd, P. and Whiteley, P. (2004) *Citizenship in Britain: Values, Participation and Democracy*. Cambridge: Cambridge University Press.

Peck, J. (2011) 'Creative moments: working culture, through municipal socialism and neoliberal urbanism', in E. McCann and K. Ward (eds), *Mobile Urbanism: Cities and Policymaking in the Global Age*. Minneapolis: University of Minnesota Press.

Peck, J. and Theodore, N. (2007) 'Variegated capitalism', *Progress in Human Geography* 31 (6): 731–772.

Peet, R. and Watts, M. (1996) *Liberation Ecologies: Environment, Development, Social Movements*. New York: Routledge.

Penrose, J. (2002) 'Nations, states and homelands: territory and territoriality in nationalist thought', *Nations and Nationalism* 8 (3): 277–297.

Philo, C. (2008) 'The difficult works in the geographer's library: reading "theory and methods" in geography', in C. Philo (ed.), *Theory and Methods: Critical Essays in Human Geography*. Aldershot: Ashgate, pp. xiii–xliv.

Pianigiani, O. (1907) *Vocabolario etimologico della lingua italiana*. Roma: Albrighi e Segati.

Pickerill, J. (2003) *Cyberprotest: Environmental Activism On-line*. Manchester: Manchester University Press.

Pickles, J. (2004) *A History of Spaces. Cartographic Reason, Mapping and the Geo-Coded World*. London: Routledge.

Pleyers, G. (2010) *Alter-Globalization: Becoming Actors in a Global Age*. Cambridge: Polity Press.

Pounds, N. (1954) 'France and "les limites naturelles" from the seventeenth to twentieth century', *Annals of the Association of American Geographers* 44: 51–62.

Power Enquiry (2006) *Power to the People: The Report of Power: An Independent Inquiry into Britain's Democracy*. York: The POWER Inquiry.

Prigogine, I. and Stengers, I. (1984) *Order Out of Chaos: Man's New Dialogue with Nature*. New York: Bantam.

Purcell, M. (2008) *Recapturing Democracy: Neoliberalization and the Struggle for Alternative Urban Futures*. London: Routledge.

Putnam, R. (2000) *Bowling Alone: The Collapse and Revival of American Community*. London: Simon and Schuster.

Raffestin, C. (1978) 'Evoluzione storica della territorialità in Svizzera', in J.B. Racine, C. Raffestin and V. Ruffy (eds), *Territorialità e paradigms centro-periferia. La Svizzera e la Padania*. Milano: Unicopli, pp. 11–26.

Raffestin, C. (1980) *Pour une géographie du pouvoir*. Paris: LIT EC.

Raffestin, C. (1986) 'Ecogénèse territoriale et territorialité', in F. Auriac and R. Brunet (eds), *Espaces, jeux et enjeux*. Paris: Fayard, pp. 173–185.

Raffles, H. (1999) 'Local theory: nature and the making of an Amazonian place', *Cultural Anthropology* 14 (3): 324.

Raffoul, F. and Pettigrew, D. (2007) 'Translators' introduction', in J.L. Nancy (ed.), *The Creation of the World, or Globalization*. Albany: SUNY Press.

Retort Collective (2003) *Afflicted Powers*. London: Verso.

Robinson, W.I. (2008) *Latin America and Global Capitalism: A Critical Globalization Perspective*. Baltimore: Johns Hopkins University Press.

Rorty, R. (1999) *Philosophy and Social Hope*. London: Penguin.

Rothenberg, M.A. (2010) *The Excessive Subject: A New Theory of Social Change*. Cambridge: Polity Press.

Routledge, P. (2008) 'Acting in the network: ANT and the politics of generating associations', *Environment and Planning D: Society and Space* 26 (2): 199–217.

Routledge, P. (2009) 'Transnational resistance: global justice networks and spaces of convergence', *Geography Compass* 3 (9): 1881–1901.

Routledge, P. Cumbers, A. and Nativel, C. (2006) 'Entangled logics and grassroots imaginaries of global justice networks', *Environmental Politics* 15: 839–859.

Routledge, P., Cumbers, A. and Nativel, C. (2007) 'Grassrooting network imaginaries: relationality, power, and mutual solidarity in global justice networks', *Environment and Planning A* 39: 2575–2592.

Routledge, P. and Cumbers (2009) *Global Justice Networks: Geographies of Transnational Solidarity*. Manchester: Manchester University Press.

Rustin, M. (1988) 'Absolute voluntarism: critique of a post-Marxist concept of hegemony', *New German Critique* 43 (Winter): 146–173.

Rustin, M.E. (1999) 'Multiple families in mind', *Clinical Child Psychology and Psychiatry* 4 (1): 51–62.

Rustin, M.E. and Rustin, M.J. (2010) 'States of narcissism', in E. McGinley and A. Varchevker (eds), *Mourning, Depression and Narcissism throughout the Life Cycle*. London: Karnac.

Rutherford, T.D. (2009) 'Labour geography', in N.J. Thrift and R. Kitchin (eds), *International Encyclopedia of Human Geography*. Amsterdam: Elsevier, pp. 72–78.

Rutherford, T.D. (2010) 'De/re-centring work and class? A review and critique of labour geography', *Geography Compass* 2 (6): 768–777.

Rutherford, T.D. and Holmes, J. (2007) '"We simply have to do that stuff for our survival": labour, firm innovation and cluster governance in the Canadian automotive parts industry', *Antipode* 39 (1): 194–221.

Said, E.W. (1993) *Culture and Imperialism*. London: Chatto and Windus.

Sánchez, J.E. (1991) *Espacio, Economía y Sociedad*. México, España: Siglo Veintuno Editores.

Santos, M. (1978) *Por uma geografia nova: da crítica da geografia à geografia crítica*. São Paulo: Hucitec and Edusp.

Santos, M. (1996) *A natureza do espaço*. São Paulo: Hucitec.

Sassen, S. (2006) *Territory, Authority, Rights: From Medieval to Global Assemblages*. Princeton, NJ: Princeton University Press.

Savage, M. (2008) 'Histories, belongings, communities', *International Journal of Social Research Methodology* 11 (2): 151–162.

Sayer, R.A. (1982a) 'Explaining manufacturing shift: a reply to Keeble', *Environment and Planning A* 14 (1): 119–125.

Sayer, R.A. (1982b) 'Explanation in economic geography: abstraction versus generalization', *Progress in Human Geography* 6 (1): 68–88.

Sayer, R.A. (1991) 'Behind the locality debate: deconstructing geography's dualisms', *Environment and Planning A*, 23 (2): 283–308.

Schoenberger, E. (2000) 'The living wage in Baltimore: impacts and reflections', *Review of Radical Political Economics* 32 (3): 428–436.

Schoultz, L. (1999) *Beneath the United States*. London: Harvard University Press.

Secomb, L. (2003) 'Interrupting mythic community', *Cultural Studies Review* 9 (1): 85–100.

SEEDA (2006) *The Regional Economic Strategy 2006–2016: A Framework for Sustainable Prosperity*. Guildford: South East England Development Agency.

Seidman, G. (2007) *Beyond the Boycott: Labor Rights, Human Rights, and Transnational Activism*. New York: Russell Sage Foundation.

Sen, A. (2007) *Identity and Violence*. London: Penguin Books.

Sen, A. (2010) *The Idea of Justice*. London: Penguin Books.

Shapiro, M.J. (1997) *Violent Cartographies: Mapping Cultures of War*. Minneapolis: University of Minnesota Press.

Silver, B.J. (2003) *Forces of Labor: Workers' Movements and Globalization since 1870*. Cambridge: Cambridge University Press.

Sklair, L. (1995) 'Social movements and global capitalism', *Sociology* 29: 495–512.

Slater, D. (2009) 'Exporting imperial democracy: critical reflections on the US case', *Human Geography* 2 (3): 24–36.

Slater, D. (2010) 'Rethinking the imperial difference: towards an understanding of US–Latin American encounters', *Third World Quarterly* 31 (2): 185–206.

Smith, L.T. (2007) 'The native and the neoliberal down under: neoliberalism and "Endangered Authenticities"', in M. de la Cadena and O. Starn (eds), *Indigenous Experience Today*. Oxford: Berg, pp. 333–352.

Smith, N. (1984) *Uneven Development: Nature, Capital and the Production of Space*. Oxford: Blackwell.

Smith, T. (1994) *America's Mission: The United States and the Worldwide Struggle for Democracy in the Twentieth Century*. Princeton, NJ: Princeton University Press.

Soja, E. (1989) *Postmodern Geographies*. London: Verso.

Soja, E. (2010) *Seeking Spatial Justice*. Minneapolis: University of Minnesota Press.

Sorrenson, M.P.K. (1989) 'Towards a radical reinterpretation of New Zealand history: the role of the Waitangi Tribunal', in I.H. Kāwharu (ed.), *Waitangi. Maori and Pakeha Perspectives of the Treaty of Waitangi*. Oxford: Oxford University Press, pp. 158–178.

Sparke, M. (2005) *In the Space of Theory: Postfoundational Geographies of the Nation-State*. Minneapolis: University of Minnesota Press.

Staeheli, L. (2010) 'Political geography: democracy and the disorderly public', *Progress in Human Geography* 34 (1): 67–78.

Stafford, B.M. (2007) *Echo Objects: The Cognitive Work of Images*. Chicago: University of Chicago Press.

Stahler-Sholk, R., Vanden, H.E. and Kuecker, G.D. (eds) (2008) *Latin American Social Movements in the Twenty-First Century*. Boulder, CO: Rowman and Littlefield.

Steel, R. (1995) *Temptations of a Superpower*. Cambridge, MA: Harvard University Press.

Stengers, I. (1997) *Power and Invention: Situating Science*. Minneapolis: University of Minnesota Press.

Sturani, M.L. (2008) 'Le rappresentazioni cartografiche nella costruzione di identità territoriali: materiali e spunti di riflessione dalla prospettiva della storia della cartografia', in L. Blanco (ed.), *Organizzazione del potere e territorio. Contributi per una lettura storica della spazialità*. Milan: Franco Angeli, pp. 189–213.

Swyngedouw, E. (2000) 'The Marxian alternative: historical-geographical materialism and the political economy of capitalism', in E. Sheppard and T. Barnes (eds), *A Companion to Economic Geography*. Oxford: Blackwell, pp. 41–59.

Tarrow, S. (2005) *The New Transnational Activism*. Cambridge: Cambridge University Press.

Taylor, P.J. (1994) 'The state as container: territoriality in the modern world-system', *Progress in Human Geography* 18 (2): 151–162.

Thakur, R. (1995) 'In defence of multiculturalism', in S.W. Greif (ed.), *Immigration and National Identity: One People, Two Peoples, Many Peoples?* Palmerston North, New Zealand: Dunmore Press, pp. 255–281.

Thompson, D. (2007) *Pessimism of the Intellect: A History of New Left Review*. Monmouth: Merlin Press.

Thompson, P. (1987) 'A review of the Open University course: work and society', *Work, Employment and Society* 1 (1): 129–133.

Thrift, N. (2007) *Non-Representational Theory: Space, Politics, Affect*. London: Routledge.

Tormey, S. (2004) *Anti-capitalism: A Beginner's Guide*. Oxford: Oneworld.

Tufts, S. (2007) 'World cities and union renewal', *Geography Compass* 1 (3): 673–694.

Tufts, S. (2009) 'Hospitality unionism and labour market adjustment: toward Schumpeterian unionism?', *Geoforum* 40 (6): 980–990.

Tufts, S. and Savage, L. (2009) 'Labouring geography: negotiating scales, strategies and future directions', *Geoforum* 40 (6): 945–948.

Turner, S. (1999) 'Settlement as forgetting', in K. Neumann, N. Thomas and H. Ericksen (eds), *Quicksands: Foundational Histories in Australia and Aotearoa New Zealand*. Sydney: University of New South Wales Press, pp. 20–38.

Urry, J. (1986) 'Making space for space', *International Journal of Urban and Regional Research* 15 (2): 273–280.

Urry, J. (2004) 'Small worlds and the new "social physics"', *Global Networks* 4: 109–130.

USSOUTHCOM (2008) *Partnership for the Americas*. United States Southern Command: Command Strategy 2018.

Valins, O. (1999) 'Identity, space and boundaries: ultra-Orthodox Judaism in contemporary Britain'. Unpublished PhD Thesis, Department of Geographical and Earth Sciences, University of Glasgow.

Vertovec, S. (2007) *New Complexities of Cohesion in Britain: Super-Diversity, Transnationalism and Civil Integration*. London: Commission on Integration and Cohesion.

Virno, P. (2004) *A Grammar of the Multitude: For an Analysis of Contemporary Forms of Life*, trans. I. Bertoletti and J. Cascaito. Los Angeles: Semiotext(e).

Wainwright, H. (1994) *Arguments for a New Left: Answering the Free Market Right*. Oxford: Blackwell.

Wainwright, H. (2011) 'Transformative resistance: the role of labour and trade union in alternatives to privatisation', in D. MacDonald and G. Reutters (eds), *Alternatives to Privatisation in the Global South*. London: Routledge.

Wainwright, H. and Elliott, D. (1982) *The Lucas Story: A New Trade Unionism in the Makings?* London: Allison and Busby.

Wainwright, H. and Little, M. (2009) *Public Service Reform But Not as You Know It*. London: Picnic Books with Compass and UNISON.

Walker, R. (1982) 'A lot to learn about time'. Wellington: *NZ Listener*.

Walker, R. (1999) 'Putting capital in its place: globalization and the prospects for labor', *Geoforum* 30 (3): 263–284.

Wallerstein, I. (1997a) 'El espaciotiempo como base del conocimiento', *Análisis Político* 32.

Wallerstein, I. (1997b) 'The time of space and the space of time: the future of social sicence', *Political Geography* 17 (1): 71–82.

Walsh, J. (2000) 'Organizing the scale of labor regulation in the United States: service-sector activism in the city', *Environment and Planning A* 32 (9): 1593–1610.

Ward, K. (2007) 'Thinking geographically about work employment and society', *Work, Employment and Society* 21 (2): 265–276.

Warde, A. (1985) 'Spatial change, politics and the division of labour', in D. Gregory and J. Urry (eds), *Social Relations and Spatial Structures*. London: Macmillan, pp. 190–212.

Waterman, P. and Wills, J. (eds) (2001) *Place, Space and the New Labour Internationalisms*. Oxford: Blackwell.

Watson, S. (1999) 'City politics', in S. Pile, C. Brook and G. Mooney (ed.), *Unruly Cities*. London: Routledge, pp. 201–246.

Whatmore, S. (2002) *Hybrid Geographies: Natures, Cultures, Spaces*. London: Routledge.

Williams, R. (1961) *The Long Revolution*. New York: Columbia University Press.

Williams, R. (1979) *Politics and Letters: Interviews with New Left Review*. London: Verso.

Williams, W.A. (2007) *Empire as a Way of Life*. New York: IG Publishing.

Wills, J. (2002) 'Political economy III: neoliberal chickens, Seattle and geography', *Progress in Human Geography* 26 (1): 90–100.

Wills, J. (2004) 'Campaigning for low paid workers: the East London Communities Organisation (TELCO) Living Wage Campaign', in G. Healy, E. Heery, P. Taylor and W. Brown (eds), *The Future of Worker Representation*. Oxford: Oxford University Press, pp. 264–282.

Wills, J. (2009a) 'Subcontracted employment and its challenge to labor', *Labor Studies Journal* 34 (4): 441–460.

Wills, J. (2009b) 'The living wage', *Soundings: A Journal of Politics and Culture* 42: 33–46.

Wills, J. (2010) 'Identity making for action: the example of London Citizens', in M. Wetherell (ed.), *Theorizing Identities and Social Action*. Basingstoke: Palgrave Macmillan, pp. 157–176.

Wills, J. (2012) 'The geography of community and political organisation in London today', *Political Geography* 31: 114–126.

Wills, J., Datta, K. Evans, J. and Herbert, J. (2009) *Global Cities at Work: New Migrant Divisions of Labour*. London: Pluto.

Wilson, W. (1902) 'The ideals of America', *Atlantic Monthly* (December). At www.theatlantic.com/past/docs/issues/02dec/wilson.htm (last accessed 25 July 2012).

Winnicott, D. (1966) *The Family and Individual Development*. New York: Basic Books.

Wright, P. (1993) 'Beastly troubles of the last Politburo', *Guardian* 17 July 1993.

Zakaria, F. (2008) *The Post-American World*. London: Allen Lane.

Zinn, H. (1996) *A People's History of the United States*. London: Longman.

Index

References to notes are indicated by a lower-case n after the page reference (e.g., 15n). Titles of works by Doreen Massey are given with the date, e.g., *World City* (2007); references for works by other authors also include the authors' names, e.g., *New Reflections on the Revolution of Our Time* (Laclau, 1990).

Spatial Politics: Essays for Doreen Massey, First Edition.
Edited by David Featherstone and Joe Painter.
© 2013 John Wiley & Sons, Ltd. Published 2013 by John Wiley & Sons, Ltd.

Printed and bound by CPI Group (UK) Ltd, Croydon, CR0 4YY

27/10/2024

14580191-0002